M.G. Sumithra, Rajesh Kumar Dhanaraj, Celestine Iwendi
and Anto Merline Manoharan (Eds.)
Deep Learning for Cognitive Computing Systems

De Gruyter Series on Smart Computing Applications

Edited by
Prasenjit Chatterjee, Dilbagh Panchal, Dragan Pamucar,
Sarfaraz Hashemkhani Zolfani

Volume 7

Deep Learning for Cognitive Computing Systems

Technological Advancements and Applications

Edited by
M.G. Sumithra, Rajesh Kumar Dhanaraj,
Celestine Iwendi and Anto Merline Manoharan

DE GRUYTER

Editors
Dr. M.G. Sumithra
Professor & Head, Department of BME (R&D),
Dr. NGP Institute of Technology,
Coimbatore - 641048
Tamil Nadu, India
sumithrapalanisamy74@gmail.com

Dr. Celestine Iwendi
Bangor College China
Bangor
Gwynedd LL57 2DG
Great Britain
celestine.iwendi@ieee.org

Dr. Rajesh Kumar Dhanaraj
Galgotias University
Yamuna Expy
Plot No. 2
Greater Noida 203201
Uttar Pradesh
India
sangeraje@gmail.com

Anto Merline Manoharan
Researcher, Department of I & C,
Dr. NGP Institute of Technology,
Coimbatore - 641048
Tamil Nadu, India
merlinemanoharan@gmail.com

ISBN 978-3-11-075050-8
e-ISBN (PDF) 978-3-11-075058-4
e-ISBN (EPUB) 978-3-11-075061-4
ISSN 2700-6239

Library of Congress Control Number: 2022942859

Bibliographic information published by the Deutsche Nationalbibliothek
The Deutsche Nationalbibliothek lists this publication in the Deutsche Nationalbibliografie; detailed bibliographic data are available on the Internet at http://dnb.dnb.de.

Cover image: monsitj/iStock/Getty Images Plus
Typesetting: Integra Software Services Pvt. Ltd.
Printing and binding: CPI books GmbH, Leck

MIX
Papier | Fördert gute Waldnutzung
FSC® C083411

www.degruyter.com

Contents

List of Authors

Jahfar Ali
Language Technologies Research Centre, IIIT Hyderabad, Telangana, India,
e-mail: jahfar2004@gmail.com

Dr. R. Babu
Department of Computational Intelligence, SRM Institute of Science and Technology,
Kattankulathur Campus, Chennai, e-mail: babu.rajen17@gmail.com

K. Balasamy
Assistant Professor, Department of AI and DS, Bannari amman Institute of Technology,
Sathyamangalam, Erode, Tamil Nadu, India, e-mail: balasamyk@gmail.com

Nitesh Bharti
SRM University-AP, Andhra Pradesh, India, e-mail: nitesh_bharti@srmap.edu.in

Dineshnath G
Department of Computer Science Engineering, PEC, Pondicherry, India
e-mail: dineshnathg@pec.edu

R. Gayathri
Department of Electronics and Communication Engineering, Coimbatore Institute of Engineering
and Technology, Thondamuthur, Coimbatore, Tamil Nadu, India - 641109.
e-mail: gayathriyaso@gmail.com

Shahab Nadeem Hashmi
SRM University-AP, Andhra Pradesh, India, e-mail: shahab_nadeem@srmap.edu.in

T Jagadesh
Department of Electronics and Communication Engineering, KPR Institute of Engineering and
Technology, Arasur, Coimbatore, Tamil Nadu, India - 641407. e-mail: jackshree@gmail.com

Dr. K. Jayashree
Department of Artificial Intelligence and Data Science, Panimalar Engineering College, Poonamalle,
Chennai, e-mail: k.jayashri@gmail.com

Jothishree
Student, Sri Sairam Engineering College, Sai Leo Nagar, West Tambaram, Chennai 600 044,
Tamil Nadu, India, email: sec20it019@sairamtap.edu.in

K. Kandasamy
Assistant professor, Department of ECE, Anna University Regional Campus, Coimbatore,
Tamil Nadu, India, email: kkandasamy.03@rediffmail.com

N. Khadirkumar
Assistant Professor, Computer Science and Engineering, Maha Barathi Engineering College, Salem
Main Road, A. Vasudevanur, Chinnasalem 606201, Tamil Nadu, India, e-mail: khadir_n@yahoo.co.in

Dr. M. Krishnamoorthi
Professor, Dr. N.G.P. Institute of Technology, Dr. N.G.P. Nagar, Kalapatti Main Rd, Coimbatore
641048, Tamil Nadu, India, email: hodit@drngpit.ac.in

https://doi.org/10.1515/9783110750584-203

Pravin Kumar
Student, Sri Sairam Engineering College, Sai Leo Nagar, West Tambaram, Chennai 600 044, Tamil Nadu, India, email: sec20it076@sairamtap.edu.in

B. Lakshmanan
Department of Computer Science and Engineering, Mepco Schlenk Engineering College (Autonomous), Sivakasi 626005, Tamil Nadu, India, e-mail: lakshmanan@mepcoeng.ac.in

V. M. Manikandan
SRM University-AP, Neeru Konda, Amaravati, Andhra Pradesh, India, e-mail: manikandan.v@srmap.edu.in

Anto Merline Manoharan
Ph.D Research Scholar, Department of I&C, Anna University, Chennai, Tamil Nadu, India, e-mail: merlinemanoharan@gmail.com

Oleeviya Babu P
Otto von Guericke University, Magdeburg, Germany, e-mail: oleeviyazehr@gmail.com

S. Priyadharsini
Department of Computer Science and Engineering, Mepco Schlenk Engineering College (Autonomous), Sivakasi 626005, Tamil Nadu, India, e-mail: priyadharsini@mepcoeng.ac.in

G. Priyanka
Department of Computer Science and Engineering, Mepco Schlenk Engineering College (Autonomous), Sivakasi 626005, Tamil Nadu, India, e-mail: priyanka@mepcoeng.ac.in

Dr. P. Rajaram
Professor, Information Technology, Malla Reddy Institute of Engineering and Technology, Hyderabad, Telangana, India, e-mail: dr.rajaram@mriet.ac.in

R. Ranjana
Associate Professor, Department of Information Technology, Sri Sairam Engineering College, Sai Leo Nagar, West Tambaram, Chennai 600 044, Tamil Nadu, India e-mail: ranjana.it@sairam.edu.in

A Reethika
Department of Electronics and Communication Engineering, KPR Institute of Engineering and Technology, Coimbatore, Tamil Nadu, India, e-mail: reethikaanandan@gmail.com

Dr. R. Regan
Assistant Professor, Computer Science and Engineering, University College of Engineering Villupuram, Kakuppam, Villupuram, Tamil Nadu, India, e-mail: reganr85@gmail.com

Dr. R. Remya
Assistant Professor, Dr. N.G.P. Institute of Technology, Dr. N.G.P. Nagar, Kalapatti Main Rd, Coimbatore 641048, Tamil Nadu, India, e-mail: remiamernath@gmail.com

K. Renuka Devi
Assistant Professor, Department of Information Technology, Dr. Mahalingam College of Engineering and Technology, Pollachi, Tamil Nadu, India, e-mail: krenukagiri@gmail.com

Dr. Prithi Samuel
Department of Computational Intelligence, SRM Institute of Science and Technology, Kattankulathur Campus, Chennai, e-mail: drprithisamuel@gmail.com

Dr. S. Saraswathi
Department of Information Technology, PEC, Pondicherry, India, e-mail: swathi@pec.edu

K. Senthilkumar
Associate Professor, Information Technology, Malla Reddy Institute of Engineering and Technology, Hyderabad, Telangana, India, e-mail: kksent@gmail.com

M S Kanivarshini
HCL Technologies, 71/A, Chokaramman nagar, Erode, Tamil Nadu, India, 638301. e-mail: kanivardhinis@hcl.com

Dr. Sobin C. C
Associate Professor, Department of CSE, SRM University AP, Neeru Konda, Amaravati, Andhra Pradesh, India, e-mail: sobin.c@srmap.edu.in

Subha T
Assistant Professor, Department of Computational Intelligence, SRM Institute of Science and Technology, Chennai, Tamil Nadu, India, e-mail: subharajan@gmail.com

S. Suganyadevi
Assistant Professor, Department of ECE, KPR Institute of Engineering and Technology, Coimbatore, Tamil Nadu, India, e-mail: suganya3223@gmail.com

Dr. M.G. Sumithra
Professor, Dr. N.G.P. Institute of Technology, Dr. N.G.P. Nagar, Kalapatti Main Rd, Coimbatore 641048, Tamil Nadu, India, e-mail: Sumithrapalanisamy74@gmail.com

Varsha
Student, Sri Sairam Engineering College, Sai Leo Nagar, West Tambaram, Chennai 600 044, Tamil Nadu, India, email: sec20it111@sairamtap.edu.in

R. Regan, Anto Merline Manoharan, R. Gayathri, K. Kandasamy

The Learning of Deep Learning: Overview, Methods, and Applications

Abstract: An extraordinary self-driving car was introduced onto the busy roads of United States America a few years ago. The look of the vehicle is like all other tonomous cars demonstrated by Tesla, General Motors, or Google, but the magic of artificial intelligence (AI) was introduced in it. The car was designed in such a way that it would not follow any instruction of an engineer or programmer. But, the function of the car was taught by itself by an algorithm that was designed by watching how a human driver would do. Getting a car that performs all the functions with the magic of AI is a remarkable achievement. But it is not absolutely understandable the way the car makes its decisions. The information is received from the car's sensors and the received information has been passed directly into a huge network of artificial neurons to perform certain functions by processing the data. The response gives the impression as it would come from the human driver. That is, the magical show of AI.

With AI and deep learning, it is made possible to ask questions to a machine and get answers about stock, customer relation, sales, fault detection, and much more. The computer can also determine to provide relevant information that is actually asked. AI provides a brief summary of the data and suggests the possible ways to analyze it. In health care field, a firm decision is taken on the efficiency of the treatment and plug-in or supplementary items are easily recommended at a greater rate in retail applications. In finance department, fault can be stopped from happening instead of just getting attention. In above exemplified applications, the physical system easily recognizes the needed information, tries to find the relationships among the used variables, and formulates an answer. Once the answer is formulated, the system will automatically communicate with options for follow-up queries.

Keywords: Deep learning, artificial intelligence, machine learning, sparse tensor network engine

R. Regan, Department of CSE, University College of Engineering, Villupuram
Anto Merline Manoharan, Department of I&C, Anna University, Chennai
R. Gayathri, Department of ECE, CIET College of Technology, Coimbatore
K. Kandasamy, Department of ECE, Anna University Regional Campus, Coimbatore

https://doi.org/10.1515/9783110750584-001

Introduction

The area of artificial intelligence (AI) is the study of how machines execute the tasks which could be completed with the help of human brilliance or intelligence [1, 2]. AI is an advanced area of computing technology that enables a computer or a machine's critical thinking ability as an intelligent person's brain. AI can be defined as a collective study of the processes by which the human brain reads the problem, gets knowledge on problems, finds the ways to solve the problems, takes decisions to find the best way, and executes the tasks to achieve the solution.

AI's major goal is to improve computing's functions to the level of a human brain by adding activities like thinking, learning, and problem solving [3]. The intelligence is difficult to describe or measure and it is capable of doing the functions like reasoning, learning, problem solving, perception, and linguistic intelligence [4].

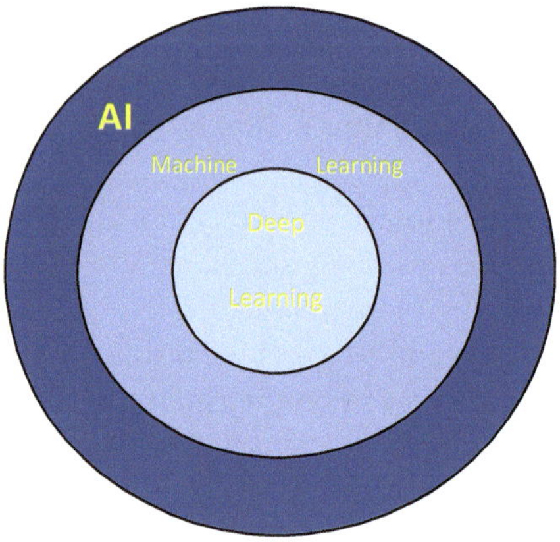

Figure 1: AI versus machine leaning versus deep learning.

In Figure 1, the AI has extended its service in various fields by making certain magic to happen in military processes and warfare, numerical data analysis, and science of human brain with psychology and philosophy [5]. AI is the field in which more attention is given by many researchers to bring about the computers significantly helpful and competent of independent reasoning to a greater extent [6, 7]. The AI is subdivided into different disciplines:

Machine learning makes use of various methods from science and statistical studies such as artificial neural networks, physics, numerical data analysis, and analytical study of problem-solving and decision-making to understand concealed facts in data without any explicit instruction where to search or what to deal [8]. *Artificial neural network* is a similar style of machine learning that is stimulated by a person's brain's activities. In neural network, the system is made up of interdependent units (as neurons in brain) that are responsible to manipulate information by reacting to external inputs [9]. *Deep learning* uses a large neural network in which the data is processed through multiple layers. Deep learning becomes much feasible as the computing power and training techniques are much improved that makes learning the difficult patterns much easier when dealing with huge datasets. Deep learning refers to a computing idea by which the machines can perform the required tasks smartly [10, 11].

Machine Learning

AI contains the subset in terms of machine learning in which the machines can understand the patterns by experience and develop skills with lack of any individual participation [12]. Machine learning is defined as a subdivision of AI, and machine learning approaches train the machines to support and take advantages from learnt experience and perform the necessary functions without being explicitly given the instructions to do so. So, it is easy to define the machine learning as the practice of making machines to solve problems by acquisition of the ability to think [13].

Deep Learning

It refers to subspecialty area of machine learning. Deep learning is designed and aided by guiding algorithms which are developed on human brain's functions and structure [14]. Deep learning is a type of machine learning in which more emphasis is placed on educating the machine or computer to understand the underlying nature of humans [15].

Deep learning is a form of machine learning method that looks at things from the ground up. It trains a computer to sort out and filter the incoming data using layers in order to learn how to forecast and analyze the information [16]. The images, text, or sound forms are used to express the observations.

The human brain and the way it filters information is the driving force behind deep learning [17]. Deep learning's major objective is to teach a computer to think and do the magic in the same way as the human brain does.

Deep learning is a subfield of machine learning and AI, as previously stated that acts in the same manner as how humans acquire certain types of information [18]. Deep learning relies on the numerical data modeling and predictive modeling that are considered as major areas of data science [19]. The primary functions of data science are collection, analysis, and interpretation of huge dataset. Deep learning strategy helps the data scientists to perform such functions faster and easier [20].

To understand deep learning better, it can be considered as one of the best techniques to computerize predictive analysis automatically. Because machine learning algorithms are linear in nature and deep learning algorithms are a subset of machine language, deep learning algorithms are loaded in increasing order of complexity and level of abstraction. For example, consider the scenario which describes how a child understands a *cat*. By pointing to various bits and pieces of a cat and understanding the word "cat," the child learns what a cat is and is not [21]. The father tries to make the child to understand "what a cat is or is not" by explaining various objects of a cat. As the child starts to spot the objects, he develops the knowledge of the objects that the cat really holds. The child understands the concept of a cat without knowing it. It is an example of difficult level of abstraction by establishing a hierarchical structure of understanding in which each level is built on the previous layer's gained knowledge or information. The cat is recognized by deep learning and machine learning algorithms in Figure 2.

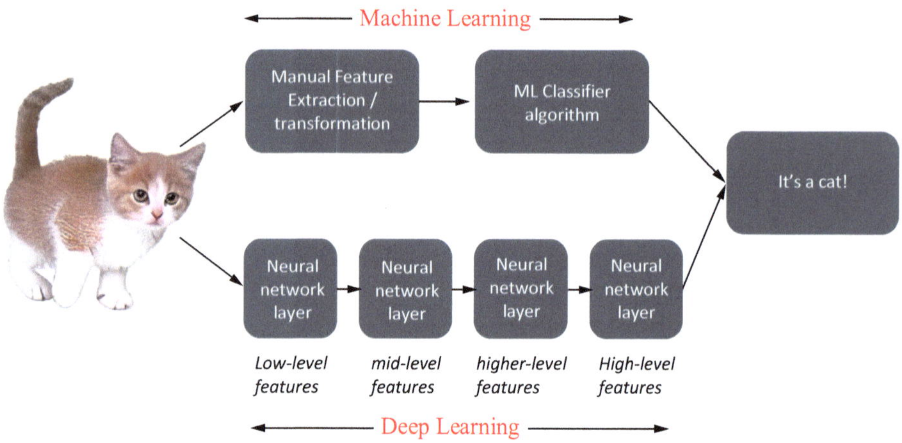

Figure 2: Deep learning versus machine learning recognition.

Deep learning is able to consume and assess more data than ever before thanks to high-powered computer processing and massive data storage. High-performance devices are connected to the Internet to produce huge dataset. AI achieves its key objective of developing smart machines thanks to breakthroughs or advancements in computing technology and the arrival of the internet. People have begun to incorporate deep learning into their day-to-day activities. AI and deep learning power a wide range of Internet services, from retail buying sites to social media autogenerated photo captions.

Deep Learning Versus Machine Learning

In spite of the fact that machine learning has a subset called deep learning, there are many factors to make the difference between them.

Machine learning is a ground of AI that allows computers to absorb from a range of datasets and make predictions based on their previous experiences, whereas deep learning analyzes the entire universe as a layered hierarchy of concepts in a more flexible way. Table 1 describes comparison of Deep learning and Machine Learning.

Table 1: Comparison of deep learning and machine learning.

Detail	Deep learning	Machine learning
Mechanism of operation	The characteristics of data and their correlations are comprehended in deep learning, and the data are processed in many phases.	Machine learning makes use of automated algorithms to forecast future results, and functions are processed built on the data provided to it.
Management	All the deep learning algorithms are self-sufficient or autonomous which are executed to get the desired result and data analysis.	All deep learning algorithms are self-contained or autonomous, and they are used to produce the desired outcome and data analysis.
Practical examples	Deep learning is useful in a variety of applications such as virtual assistants, shopping and entertainment, facial recognition, translation, medical care, and self-driving car vision.	One of the most popular uses of machine learning is image recognition. It is used to identify hings like people, locations, and digital photographs. Medical diagnostics, classification, prediction, and extraction are all examples of machine learning applications.
Data points	Millions of data pieces are analyzed in deep learning.	In machine learning, just a few data points, usually thousands, are needed for analysis.
Time for training	Deep learning algorithms require greater parameters and take a lengthy time to train.	In general, machine learning algorithms require less time to train and the algorithms normally consume a few minutes to hours of time for training.
Output	In deep learning, the output can be miscellaneous and it could be in the form of a classification, a score, an element, or even just a text.	The result of machine learning is usually a numerical classification value.

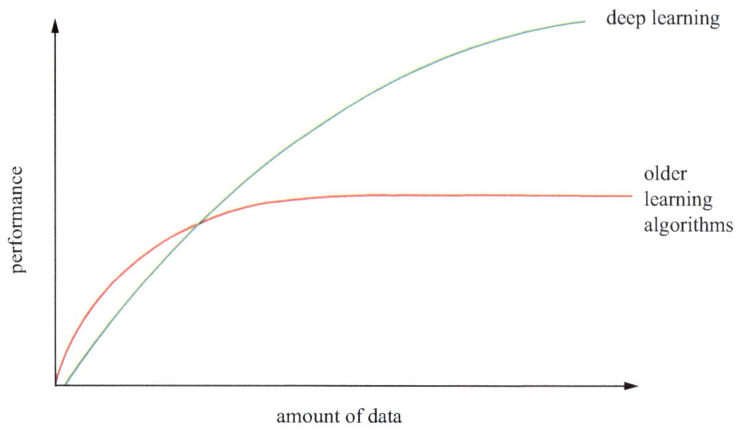

Figure 3: Performance comparison between deep learning and others.

The main difference between deep learning and traditional machine learning is how effectively it scales up with data. In Figure 3, deep learning techniques do not perform well when the data is minimal as deep learning algorithms need a substantial quantity of data to fully comprehend it. But the standard machine learning algorithms can perform with their customized rules and dominate in the limited data case [22]. In Figure 4, deep learning performs better with the large dataset comparing to machine learning.

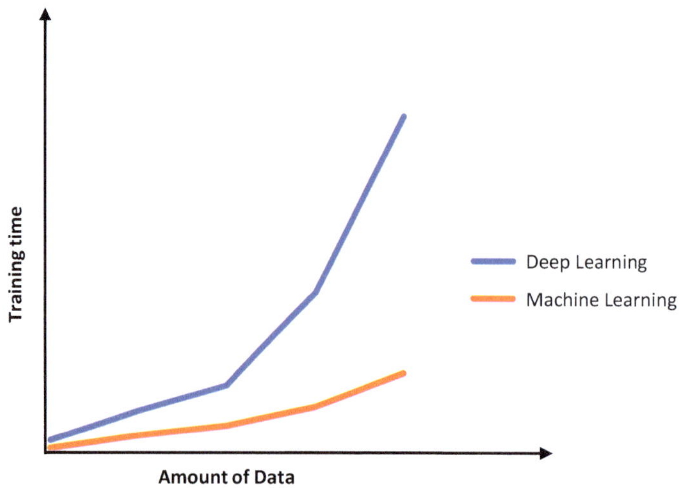

Figure 4: Training time comparison of deep learning with machine learning.

A deep learning algorithm demands a prolonged time to train because it has numerous factors that allow it to take longer time to train than a traditional algorithm. ResNet, a cutting-edge deep learning system, takes approximately several weeks of training time from the start. Machine learning, on the other hand, takes less time to train, approximately a few hours.

How Deep Learning Works

In deep learning, each algorithm in the hierarchy gets the input and executes it. The input is processed and learnt to generate a statistical model in the form of an output. The transformation process must be continued as far as the output has attained an appropriate level of precision [23].

In deep learning, the term "deep" refers to the number of processing stages in a hierarchy that data must pass through to achieve the intended result. To begin with the deep learning process, the computer program accepts the training data which may be in the form of a set of images. Each image is labeled and represented by meta tags. For example, the computer takes what it learns from the training data and creates a feature set for "cat." The process is further continued to build a predictive algorithm that predicts that anything in an image with four legs and a tail should be classified as a cat. The program will search for pixels patterns data from digital images. The predictive algorithmic model is processed numerous times until the desired level of accuracy is reached, and the model turns out to be more difficult and precise for each level of processing iteration.

The child actually takes a few weeks, if not a month, to grasp the perception of *a cat*, but a deep learning based computer program accurately recognizes which images contain cats in them within a few minutes [24]. Deep learning achieves the short response time as the deep learning algorithm is allowed to sort millions of images of a cat and train with the large dataset.

To achieve an affordable level of accuracy, deep learning techniques utilize large datasets and considerable computing power. Prior to the arrival of large dataset and cloud computing, both datasets and processing capacity were extremely difficult to come by. The data created by both the machines as well as humans is unlabeled and disorganized. As deep learning can use its own output to build complicated numerical statistical models through iterations, it becomes very simple to build précised prediction model seven for large amount of unregulated and unlabeled data.

As the human brain has a large number of neurons, neural networks are built up of nodes arranged in layers. Individual layer nodes are connected to nodes in neighboring layers. The term "deep" refers to the sum of layers in a neural network. In the human brain, a single neuron can receive millions of signals from other neurons. The impulse signals are transferred between the neural network nodes in the same manner as the human brain and the weights are tagged correspondingly. A node with a higher weight will exert more effort on the nodes below it. The weighted inputs are accumulated in the final layer to produce an output. Deep learning systems remain in the need of powerful machines as the systems need to process a large set of data and involve in several complex mathematical calculations [13].

Deep learning systems need to process huge dataset to generate more precise output. In deep learning, the data is processed in order to organize it using classification based on the responses to a sequence of binary value producing true or false questions. The data processing also involves exceptionally difficult mathematical-based calculations. For instance, a facial recognition system executes the data processing by learning experience to recognize the lines, curves and edges of the face, other important parts of the face, and the representation of face on the whole. The program educates itself by better training with each iteration, and the level of accuracy rises with the level of difficulty and effort.

In deep learning, a specific algorithm is used to learn the composite data that exists in the form of digital images, simple text, or even sound, and deep learning also categorizes the functions to be executed on composite data [18]. The developed algorithms are capable of achieving cutting-edge accuracy and they can even produce better results than human-level performance in some cases.

Deep learning is ready to lend a hand to design the machines or computers by neural network architectural layers and train the machines with large set of labeled data. Deep learning is an emerging technology which plays a vital role in different claims such as driverless robot cars, facial recognition system, and virtual assistants. The idea behind these applications is to train the computers with the data and make them learn from the past experiences.

"Deep" is the word that describes the learning ability of the neural networks that are capable of noticing and realizing the new levels of data as quickly as possible for every passing minute. Whenever new level of data is noticed, appropriate training is given with the focus of improving the performance. As the depth of the data increase, the data experts adopt high-performance training and improved level of deep learning competence accordingly. Along with the numerous advantages, there are also risks associated with deep learning undiscovered abilities.

There are different types of learning, such as supervised, semisupervised, and unsupervised learning, used to provide different levels of training according to various representations of data.

Deep learning's functionality is dependent on the following factors:

- Deep learning emulates the functionalities of a brain of human which normally involves supervising the data and generating the patterns that are used in making decisions.
- The well-trained dataset can be interrelated to each other or unconnected. The trained dataset can be naturally complicated and composite.

- When the dataset is too large, the more reliable and effective training has to be adopted to achieve better performance in the decision making Figure 5 represents model of deep learning. Deep learning is associated with bounded size layers and the layers can be of heterogeneous kind and they are different from biologically informed connectionist models to achieve efficiency, trainability, and understandability.

Deep Learning Model

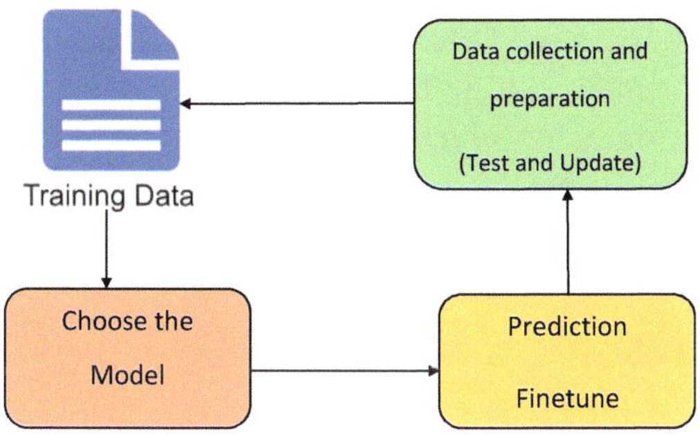

Figure 5: Deep learning model.

Obtaining Information

The first crucial stage in deep learning is simply called as data collection phase. The phase is critical because the quality and volume of data you collect will directly affect how effective the prediction model is.

Preparation of Data

The collected data have to be moved on to the next stage of deep learning or the required data are imported into an appropriate location and prepared for use in deep learning training during data preparation.

Training

The appropriate model is chosen and the data get to the training, which is typically regarded the most difficult part of deep learning. The weights (W) and the biases (b) are set. In this phase, the data is utilized to enhance our model's ability to forecast the outcome progressively. The deep learning procedure demonstrates that you must first go through a training phase. This is the stage where you effectively train the programs to produce the desired results. The training parameters are changed by the program during training. It also changes the input data before producing an output. Once an output is received, it has to be assessed. If the result is not acceptable, then the training has to be continued.

Parameter Turning

By tweaking the settings, you may optimize the training even further. There may have been some implicit assumptions made during training, and now is a good time to test those assumptions and try different values, which can sometimes lead to higher accuracy.

Deep Learning Techniques

There are several kinds of deep learning techniques that can accurately and efficiently solve issues that are extremely hard for brain of human to solve. Convolutional neural networks (CNNs), deep belief networks (DBNs), auto encoders models are some of deep learning models that are discussed in detail as follows.

Convolutional Neural Networks

CNN has several layers of artificial neurons. Pooling layers, convolutional network layer, and fully connected layers are three types of neural layers that make up a CNN. Each layer is used for a distinct purpose. Each layer of CNN converts the input to a neuron activation output, likely contributing to fully connected layers that translate the participation data (input) to a 1D feature vector. Face recognition, object identification, robotic vision, and self-driving vehicles are just a few of the applications where CNNs have proved highly effective.

Artificial neurons are mathematical models that compute the weighted summation value of numerous inputs and create an activation value, which is comparable to the output of biological neurons. Each layer of a CNN creates a number of activation functions that are passed on to the next layer when an image is received into it. Basic characteristics such as horizontal or diagonal edges are generally extracted by the first layer. The data are then passed on to the next layer, which is in charge of identifying more complex characteristics such as corners and combinational edges. CNN is becoming increasingly capable of recognizing increasingly complex situations.

CNNs are mostly employed in image processing. It applies weights and preferences to distinct items in the image and distinguishes them. In comparison to other classification methods, it requires less preprocessing time. To capture the temporal and geographic relationships in image data, CNN employs suitable filters.

Deep Belief Networks

DBNs are deep learning models that use the restricted Boltzmann machine (RBM) as a learning module, classifying them as part of the "Boltzmann family." The RBM is a reproductive neural network with directed connections to the lower layers and undirected connections to the top two levels. A DBN's initial two layers create an undirected graph, but the layers beyond that procedure a directed graph.

DBN is a strong reproductive model that employs a deep architecture consisting of several loads of RBM. Each RBM model achieves a nonlinear adjustment on its input vectors and generates output vectors. It will be utilized as input for the following RBM model in the system. As a result, DBNs gain a lot of flexibility and are easier to extend. DBNs may be utilized in either an unsupervised or supervised context because they are generative models. DBNs can learn/extract features and classify them, which are useful in a variety of applications.

The use of features for learning in the first layer is the very first phase in the DBN's training. The activation of learnt characteristics is then implemented in the following layer. The procedure is repeated until the final layer is reached. The DBNs, which are trained using RBMs, are fine-tuned using the feed-forward network. DBN learns all of the hidden patterns at once.

Autoencoder/Stack Autoencoder

The three layers (input, hidden, and output) make up an autoencoder, which is a sort of unsupervised learning model. The autoencoder training procedure has two components: an encoder and a decoder. An encoder plans the input data into a hidden representation, while a decoder rebuilds the input data from the unknown representation.

Identifying uniqueness and diagnosing faults are two challenges that need to be resolved by autoencoders. An autoencoder's one layer is intended as an encoding layer, while the other layer is designed as a decoder. The training is then supplied utilizing the unsupervised approach to preprocess the input. Once the training is over, the weights are fixed. Until the processed input is passed on to the next layer, all the layers are intended for pre-training. Then, the classification/regression step is followed and stochastic gradient descent is used for optimization. Stacked autoencoders employ the autoencoder as its primary building block, same to how RBMs are used in DBN.

In terms of the number of properties, CNNs, DBNs, and autoencoder are compared. Table 2 depicts a comparative study of deep learning approaches, with ✓ indicating high performance and X indicating poor performance or complete absence thereof. Table 3 describes advantage and disadvantage of deep learning.

Table 2: Comparison of CNN, DBN, and autoencoder.

Model properties	CNNs	DBNs	Autoencoder
Unsupervised learning	X	✓	✓
Training efficiency	X	X	✓
Feature learning	✓	X	X
Scale/rotation/translation invariance	✓	X	X
Generalization	✓	✓	✓

Deep Learning's Advantages and Disadvantages

Table 3: Advantages and disadvantages of deep learning.

Advantages	Disadvantages
It is quite common that the training dataset is limited sometimes. Deep learning, on the other hand, has the potential to create significant updates from the restricted datasets utilized for training.	In spite of increasing popularity, deep learning is also open to the threats; the security is the huge concern in deep learning that must be addressed.
Deep leaning is also handy in producing reliable task results when working on unsupervised learning techniques. When it comes to machine learning, feature engineering is one of the processes that consumes the most time. But deep learning is so flexible and useful in reducing the time taken by feature engineering.	The flow of data during the training process is continuous and uninterrupted. The phenomenon reduces the possibility of improving the training process. As the size of dataset increases, the training time also increases. So, the larger the dataset, the cost of computational training is also more.
By continuing the training constantly, deep learning makes the working architecture to be compatible to any change and the architecture becomes capable of dealing with different problems.	There is no transparency in fault revision. When a fault has occurred, there are no intermediate steps to make the changes to deal with a fault. The entire algorithm has to be revised to resolve the issue.

Table 3 (continued)

Advantages	Disadvantages
GPUs can conduct massively parallel computations, and deep learning models are scalable for enormous amounts of data. Furthermore, it provides superior outcomes while dealing with large amounts of data.	High-speed computational units and strong graphical processing units are required to achieve effective training process with the available datasets. These devices are expensive resources. So, the training process is expensive as more money has to be spent for better training.

Deep Learning in Action

It is important to know how deep learning makes an impact in people's lives. Deep learning gets its attention in the computing world by using various applications of all sizes. When new images are submitted to Facebook, the social media platform frequently classifies and tags your friends [18]. Deep learning strategy is commonly used by digital assistants like Alexa, Siri, Google, and Cortana for incorporating natural language processing feature and speech recognition ability. Another tool that uses a deep learning technique to translate spoken interactions in real time is Skype. Deep learning is used by several email providers to identify spam messages before they reach the inbox. Users may take a picture of any object using applications like CamFind, and deep learning helps digital image search technology to recognize what the thing is. PayPal has implemented deep learning to detect fraudulent payments.

Google makes good use of deep learning's advantages by incorporating it into a variety of applications and providing support. In recent times, AlphaGo computer program of Google Deepmind has defeated the long-standing champions in the Game of Go. WaveNet, developed by Deepmind, can generate speech by emulating a specific human voice with greater accuracy than other speech systems currently available in the market. Deep learning is used by the Google Translation tool to enable the features of converting voice to text and translates the text from one language to required language. Google Planet employs deep learning to pinpoint the exact location of the photo. Tensorflow is a deep learning software database established by Google that may be used to create AI applications.

Deep learning is better known only to a few people and it is in its early years. But deep learning will boom and renovate the society in the years to come. The research toward self-driving cars is growing and such cars are being tested worldwide. The sophisticated layer of neural networks in such automated cars is trained to decide which objects should be avoided; recognition of traffic lights, and adjusts the speed according to the situation. Neural networks are being used by experts of different fields to forecast the stock prices to the weather. Digital assistants with deep learning capabilities are used to forecast and get advice on whether to sell stocks or leave before a disaster. Deep learning techniques might still protect lives by detecting cancers at an early stage and incorporating them into cancer treatment plans.

Applications of Deep Learning

Virtual Assistants

The deep learning is extremely useful in recognizing and translating the language of the human's voice. Well-known virtual assistant applications include Alexa, Cortana, and Siri.

Vision for Driverless and Autonomous Cars

An autonomous car's navigation system requires human-like knowledge and competence to expose and interpret traffic scenarios, key road sign images, signal operation, pedestrians, and speed limit sign. As a result, the navigation system must cope with a significant volume of real-world data.

The algorithms' efficiency will be boosted as a consequence of the large amount of data, which will speed up the decision-making process.

Service-Oriented Chat Bots

Chatbots' continuous communication with individuals mandatorily demands better and faster responses for the purpose of extending customer service. Many tricky questions would be raised by the individual and appropriate response to be provided to react in a helpful manner. Deep learning is adopted as the better approach to extend the best training to make fitting response to difficult questions.

Translations

Deep learning supervision is a superior alternative for automatically translating voice into numerous languages. Translation software is extremely useful for visitors, travelers, and government employees who are unfamiliar with the target language.

Facial Recognition

Facial recognition is used as the security tool in tagging feature of Facebook, a social media application. Along with the social media importance, facial recognition is used to distinguish the individual based on different metrics like beard, no beard, gaining weight, weight reduction, new haircuts, and so on.

Shopping and Entertainment

The majority of online services collect personal information from users as well as their interest toward buying goods. All e-commerce-based retail and shopping applications like Alibaba and Flipkart as well as entertainment applications like Netflix and Amazon Prime video are given access to consumer's buying nature and behavior information. Individual data, such as purchasing habits, is utilized by online applications to classify and display recommendations for future purchases and viewing. The advertisements or links are sent to the individuals with a subject line "You might like to watch/buy."

Deep learning is used to make decisions about how to classify and show advertisements or links based on the individual's activity. Decision-making efficiency improves as more data is provided as input to the deep learning algorithm.

Medical Field and Pharmaceuticals

Precision medicine aims to make the best use of massive multidimensional biological information for diagnosis assessment, therapeutic intervention methods, and prognosis. The precision medicine helps the clinicians by giving the opportunity to cautiously do early interventions to confirm whether the disease needs treatment or is preventable in nature. Deep learning systems can currently estimate risk in certain cancers and cardiovascular diseases using multidimensional clinical and biological data with reasonable accuracy.

Medicines are customized based on the patient's genetics and illness condition. Deep learning has increased the possibilities for such customized medicinal applications, and several pharmaceutical corporations have begun to use them. Fraud detection system, virtual recognition system, healthcare monitoring, entertainment services, and so on are considered as familiar deep learning based applications.

Conclusion

AI is a branch of computing technology that purposes to improve a machine's capacity to think like a human brain. Machine learning is a subfield of AI in this. Computers learn from their experiences and develop expertise abilities without the need for human intervention. Deep learning is a type of machine learning that uses many layers to extract higher-level characteristics from rare data. Deep learning's major goal is to teach a machine to think and act like a human brain.

Deep learning is a subclass of machine learning, although it differs in several ways from it. The journey of deep learning from the beginning is discussed and the roadmap from AI to deep learning is explained with important milestones. Deep learning models can be created such as learning rate evaluation, transfer knowledge, training from the start, and removing the blunders. Deep learning has two sides like the coin and each with its own set of benefits and drawbacks. Deep learning is quite useful in various applications such as automated cars, Facebook, and voice recognitions system.

References

[1] L. Deng, "A tutorial survey of architectures, algorithms, and applications for deep learning," APSIPA Transactions on Signal and Information Processing, vol. 3, pp. 1–29, 2014.

[2] L. Deng, X. He, and J. Gao, "Deep stacking networks for information retrieval," in IEEE International Conference on Acoustics, Speech and Signal Processing. IEEE, pp. 3153–3157, 2013.

[3] Y. Bengio, E. Laufer, G. Alain, and J. Yosinski, "Deep generative stochastic networks trainable by backprop," in International Conference on Machine Learning. Omnipress, Vol 6, Issue 15 pp. 226–234, 2021.

[4] E. Azarkhish, D. Rossi, I. Loi, and L. Benini, "Neurostream: Scalable and energy efficient deep learning with smart memory cubes," CoRR abs/1701.06420, Vol 1, Issue 1, 2017.

[5] R. Collobert, S. Bengio, and J. Mariéthoz, "Torch: A modular machine learning software library," Idiap-RR Idiap-RR-46-. Idiap, 2002.

[6] D. Bahdanau, K. Cho, and Y. Bengio,."Neural machine translation by jointly learning to align and translate," CoRR abs/1409.0473, 2014.

[7] N. Ballas, L. Yao, C. Pal, and A. C. Courville, "Delving deeper into convolutional networks for learning video representations," CoRR abs/1511.06432, 2015.

[8] J. Berant, A. Chou, R. Frostig, and P. Liang, "Semantic parsing on freebase from questionanswer pairs," in Empirical Methods in Natural Language Processing, Vol. 2. Association for Computational Linguistics, ijcai, p. 6, 2017.

[9] L. Breiman, "Statistical modeling: The two cultures," Quality Control and Applied Statistics, vol. 48, no. 1, pp. 81–82, 2003.

[10] D. Castelvecchi, "Can we open the black box of AI?," Nature, vol. 538, no. 7623, pp. 20–23, 2016.

[11] C. Chen, A. Seff, A. Kornhauser, and J. Xiao, "Deepdriving: Learning affordance for direct perception in autonomous driving," in IEEE International Conference on Computer Vision. IEEE, pp. 2722–2730, 2015.
[12] T. Chen and C. Chefd'hotel, "Deep learning based automatic immune cell detection for immunohistochemistry images," in International Workshop on Machine Learning in Medical Imaging. Springer, pp. 17–24, 2014.
[13] T. Chen, M. Li, Y. Li, M. Lin, N. Wang, M. Wang, T. Xiao, B. Xu, C. Zhang, and Z. Zhang, "MXNet: A flexible and efficient machine learning library for heterogeneous distributed systems," CoRR abs/1512.01274, 2015.
[14] S. Chetlur, C. Woolley, P. Vandermersch, J. Cohen, J. Tran, B. Catanzaro, and E. Shelhamer, "cuDNN: Efficient primitives for deep learning," CoRR abs/1410.0759, 2014.
[15] J.-T. Chien and H.-L. Hsieh, "Nonstationary source separation using sequential and variational Bayesian learning," IEEE Transactions on Neural Networks and Learning Systems, vol. 24, no. 5, pp. 681–694, 2013.
[16] K. Cho, B. van Merrienboer, Ç. Gülçehre, D. Bahdanau, F. Bougares, H. Schwenk, and Y. Bengio, "Learning phrase representations using RNN encoder-decoder for statistical machine translation," in The Conference on Empirical Methods in Natural Language Processing, pp. 1724–1734, 2014.
[17] M. C. Choy, D. Srinivasan, and R. L. Cheu, "Neural networks for continuous online learning and control," IEEE Transactions on Neural Networks, vol. 17, no. 6, pp. 1511–1531, 2006.
[18] D. C. Cireşan, A. Giusti, L. M. Gambardella, and J. Schmidhuber, "Mitosis detection in breast cancer histology images with deep neural networks," in International Conference on Medical Image Computing and Computer-Assisted Intervention. Springer, pp. 411–418, 2013.
[19] A. Coates, B. Huval, T. Wang, D. J. Wu, A. Y. Ng, and B. Catanzaro, "Deep learning with COTS HPC systems," in International Conference on Machine Learning. Omnipress, pp. 1337–1345, 2013.
[20] G. E. Dahl, D. Yu, L. Deng, and A. Acero, "Context-dependent pre-trained deep neural networks for large-vocabulary speech recognition," IEEE Transactions on Audio, Speech, and Language Processing, vol. 20, no. 1, pp. 30–42, 2012.
[21] N. Dalal and B. Triggs. "Histograms of oriented gradients for human detection," in IEEE Conference on Computer Vision and Pattern Recognition, vol. 1. IEEE, pp. 886–893, 2005.
[22] J. Dean, G. S. Corrado, R. Monga, K. Chen, M. Devin, Q. V. Le, M. Z. Mao, M. Ranzato, A. Senior, P. Tucker, K. Yang, and A. Y. Ng, "Large scale distributed deep networks," in The 25th International Conference on Neural Information Processing Systems. Curran Associates, pp. 1223–1231, 2012.
[23] B. Dolan, C. Quirk, and C. Brockett, "Unsupervised construction of large paraphrase corpora: Exploiting massively parallel news sources," in 20th International Conference on Computational Linguistics. Association for Computational Linguistics, 2004, pp. 350.
[24] J. Donahue, L. A. Hendricks, S. Guadarrama, M. Rohrbach, S. Venugopalan, K. Saenko, and T. Darrell, "Long-term recurrent convolutional networks for visual recognition and description," in IEEE Conference on Computer Vision and Pattern Recognition. IEEE Computer Society, 2015, pp. 2625–2634.

R. Remya, M.G. Sumithra, M. Krishnamoorthi

Foundation of Cognitive Computing

Abstract: This chapter focuses on the basics of cognitive computing methodology and explains how this acts as a platform to change the data to make the machine learn. The computer performs all the processes of evaluating and executing in a quicker manner. The cognitive system attempts to mimic the activity of the human with the support of the computer. Also, it generates a solution when more data has to be added using the machines' learning capacity. Furthermore, it explains how artificial intelligence and computers are involved in cognitive computing and its applications. Moreover, it presents an outline of cognitive computing technology.

Keywords: Cognitive computing, machine learning, AI, deep learning, healthcare

Introduction

Cognitive computing makes humans get together in touch with machines. Human investigation results in a time-consuming task, so the usage of cognitive system reduces such enormous utilization of time. Brain Krzanich stated in his work that artificial intelligence (AI) has the capability to reason, sense, and act upon it depending on the trained data [1]. The cognitive system is an AI-based system, which works by grabbing the command, then illustrating the conclusion and providing an appropriate solution. Furthermore, cognitive system supports individuals in making decisions. It is analogous to the human brain as one can access any data ranging from structured data to unstructured data. The unstructured data includes text, video, image, voice, and so on.

Ragupathi stated in his work that various sources require huge data. Furthermore, it utilizes every spectrum of the scientific and academic domains [2]. Cognitive computing reduces such shortfalls through big-data analysis [3]. The computing system collects all the human thoughts and supports the system to reduce the mistakes happening every time [4]. It creates a benefit in making a decision [5]. During World War II, the computing system became necessary only for military and scientific applications. Once the transistor and microprocessor developed, it led to the creation of bulkier computers contrasted to modern-day computers. The speed and usage of computers have reached a peak pace, and it has led to the development of portable computers and smartphones [6]. Accessing a huge amount of data has been regarded as a daunting task for a human, but the cognitive system makes ease of such task.

R. Remya, Dr. N.G.P. Institute of Technology, e-mail: remiamernath@gmail.com
M.G. Sumithra, M. Krishnamoorthi, Dr. N.G.P. Institute of Technology

https://doi.org/10.1515/9783110750584-002

The cognitive system operates at various levels when compared to the state-of-art methods. It combines the data from various sources to create the exact suggestion. For this, it utilizes pattern recognition, natural language programming (NLP), data mining, speech recognition, and so on to copy the behavior of human. It has an impact on our banking, healthcare, personal lives, and so on. Nowadays, user needs their required content in a shorter period. Cognitive search carries this task to the next level. That is, it makes the human interaction with the machine natural. For example, upon considering the medical field, the usage of cognitive computing is given in Figure 1.

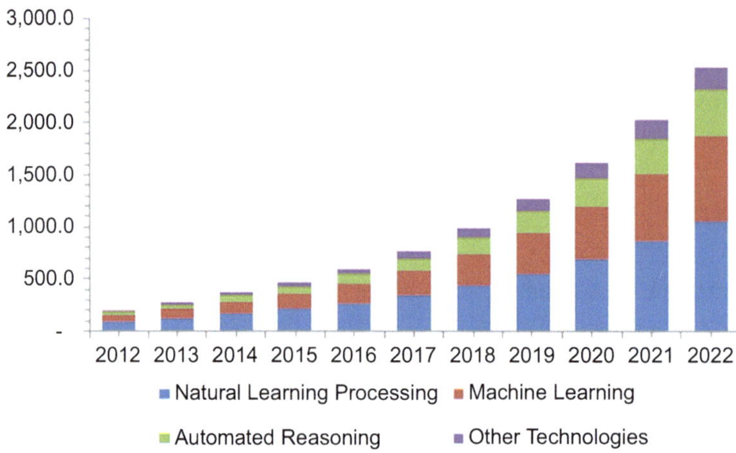

Figure 1: Cognitive computing in healthcare.

Three basic principles existing in cognitive computing include:

Learn: The machine learns any data and makes an observation on it.

Model: The system needs a model to learn the data, and assumptions has made on learning approach usage. Furthermore, context understanding is essential to how much data fits into the model.

Create hypotheses: The cognitive machine introduces that there is no single solution. The exact solution depends on the data. So, the cognitive machine is regarded as a probabilistic one. A hypothesis is one to understand the already available data. A cognitive machine utilizes the data to train such a hypothesis [7–14].

Nowadays, there is no such system available in academics or industry side. In general, cognitive computing mirrors the actions of the human brain and makes the decision. Figure 2 gives the framework of the cognitive computing system.

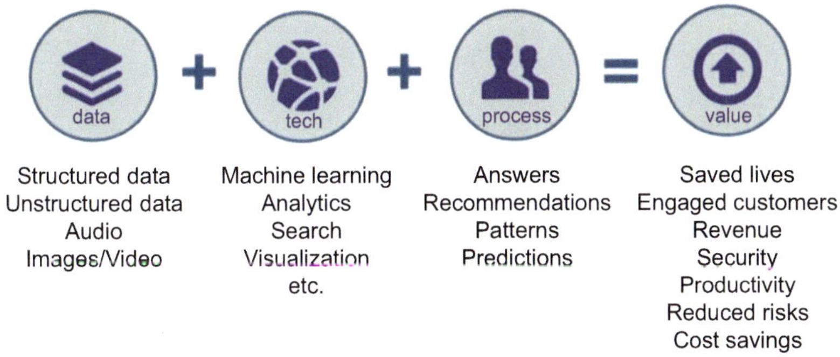

Figure 2: Cognitive computing framework.

Furthermore, it can interact with other systems to examine the data. Suitable cooperation has been necessary to make the numerous systems operate simultaneously in a real-world environment. One can communicate with the cognitive model through a human interface machine. The feature of cognitive machines has been shown in Figure 3.

Figure 3: Cognitive system features.
Source: https://www.predictiveanalyticstoday.com/what-is-cognitive-computing/ [2020]

Adaptive: Based on the requirement, aim, and data changeability, the cognitive system can learn. Furthermore, it tolerates any ambiguity. It has to construct in such a way to adapt to the changes in real-world applications.

Interactive: It has been regarded as user-friendly. It may interact with individuals, devices, and processors. Moreover, based on the previous inferences, the appropriate solution has been made to solve the problem in a real-time situation.

Iterative and stateful: The cognitive system should be aware of the previous communication and generate the information necessary for the application. It should explain the problem by raising queries or else by finding an alternative. This feature mainly focuses on the data quality to assure that the system always gives valuable information based on the up-to-date data available in the dataset.

Contextual: It extracts the contextual details such as syntax, area, and user data; the decision, job, aim, time, and so on. It has the data such as structured, constructed, and sensory data.

Cognitive computing utilizes machine learning (ML), a self-learning approach, neural networks, data mining, and visual recognition to do any human performing activities. It performs human-like activities and can solve problems. But it cannot identify emotions and spirituality.

Visual recognition utilizes neural networks and deep learning procedures to learn and identify the object given in the image or video. For example, Google lens captures the image using the phone camera and provides relevant information. NLP can be made aware of the human language. It identifies numerous counts of language and processes it. For example, while writing an email, it may suggest the next word; also, it gives an opinion about the mood of the text including formal, angry, or appreciative. In the medical field, it holds the record of patients' information, diagnostic equipment, and so on. Furthermore, it suggests the doctor to decide on the patients' health since humans do not analyze huge data regarding the patient. The main motto is to make the computer expect or behave like a human to take an appropriate decision. It leads to improved success and upgrades the knowledge.

Applications

There are so many applications available in cognitive computing; it handles tasks ranging from simple to complex. It has been performed based on logical reasoning. Some of the applications have been sorted below.

Retail Industry

It supports the marketing field to gather more information and makes retailers analyze more adaptively. Also, it has supported the companies in creating more sales and giving suggestions to the customers. E-commerce has been combined with cognitive computing to gather basic information from the customers about the product that one may be looking for. Based on the analysis, it has to recommend the customer to purchase the product [15]. The functionality of cognitive computing in the retail industry is depicted in Figure 4. Apart from this, it supports on-floor shopping. Furthermore, it will support the retailers to give personalized products and examines; what they need, when they need it, and how they need it. They derive experiences from that. It reduces waste creation by delivering fresh products to customers based on their demand.

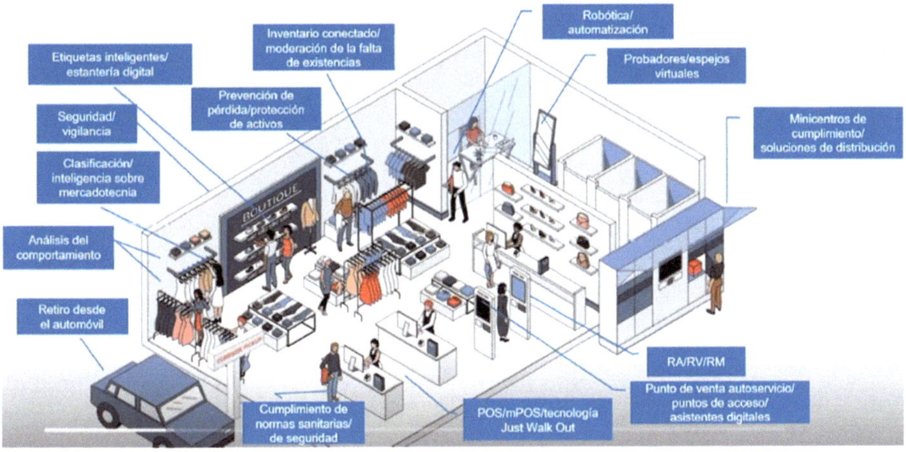

Figure 4: Cognitive computing in retail industry.

Logistics

Cognitive computing has a footprint on logistics too. It supports deciding on designing, warehousing, including management and automation, and the Internet of things (IoT). In warehousing, it supports running the storage code, computerized picking of vehicles, and usage of warehouse robots to do the work more efficiently. IoT will support warehouse management in infrastructure, improving operations in the warehouse and computerized vehicles for picking and performing the operations. In addition to IoT, another technology is wearable devices, which converts all the collected objects into sensors and makes room for numerous human decisions. It has a wide range of usage from smartwatches to ring scanners, smart glasses, smart clothes, and so on. The logistic guidance of cognitive computing is given in Figure 5.

Banking and Finance

While considering the banking industry, cognitive computing aims to improve operational efficiency, experience, and revenues. It reshapes the banking and financial organizations in three dimensions: deeper contextual engagement, new analytics insights, and enterprise information. This technology will be more supportive in the area of product management and support in customer service [16]. Cognitive banking provides support to customers and support in deciding the investment plan. Furthermore, it will be more helpful in involvement among the financial institution and the customer. In this case, the computer will understand the customers' personalities

Figure 5: Logistic guidance.
Source: https://www.facebook.com/Logistic-Guidance-109566054269122 [2020]

depending on the data available on the computer. Figure 6 depicts the use of cognitive computing in banking and finance.

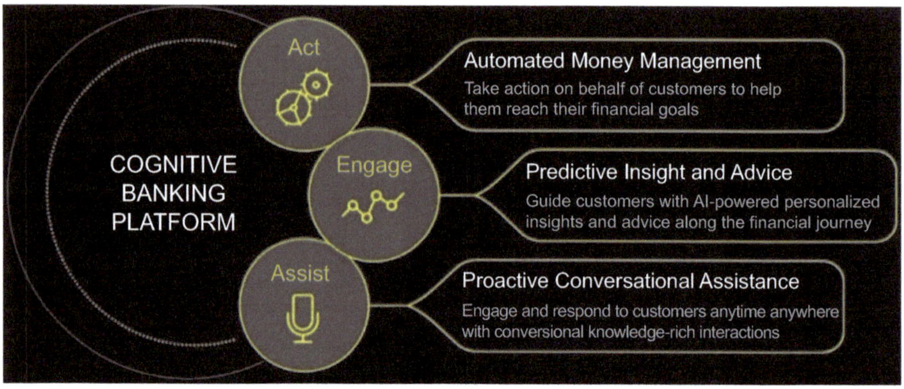

Figure 6: Cognitive computing in banking and finance.
Source: https://www.linkedin.com/pulse/power-cognitive-banking-excited-announce-our-next-customer-roberts [2018]

Cybersecurity

The cognitive procedure provides end-to-end protection and will detect threats. Besides, it will prevent hacking; and provides a solution to identify any misleading data, a rise in cyberattacks and reduces manpower in cybersecurity specialist. Such a cognitive computing system not only detects the threats, but also scans the difficulties raised in the system. Additionally, it provides a solution for it. At last, protecting the data is regarded as the most important one. To consider the advantage of cognitive computing, it needs to have a database to hold all the data. The important functionality of it is to maintain its confidentiality also to be held good [17]. The cybersecurity utilization by using cognitive computing is shown in Figure 7.

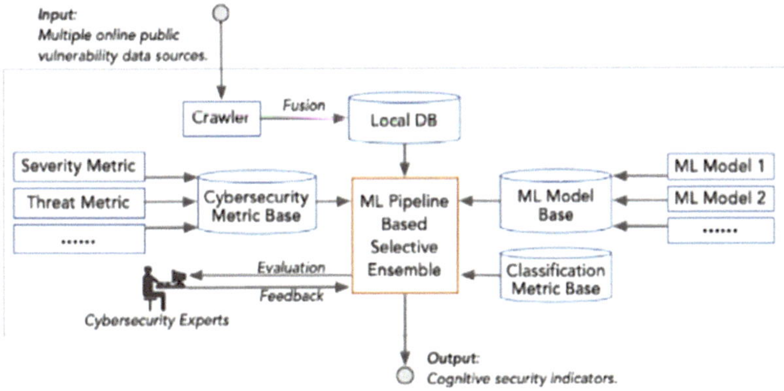

Figure 7: Cognitive computing in cybersecurity.

Healthcare

Cognitive computing supports the medical professionals to provide treatment decisions in a better way; it upgrades the efficacy of the medical experts. Also, it provides better results to the patient. It processes the data which has been relevant to the patient and provides a proper recommendation. It interconnects the functionality of the brain and computer to upgrade the decision-making capability of the healthcare professionals [18]. That is why it becomes more helpful in making an appropriate decision during the diagnosis period. Figure 8 shows the healthcare application with the support of cognitive computing.

Healthcare in treating cancer: Rare cancer poses a great threat to humans, so the cognitive system provides solutions to the extent of the caring facility and makes clinicians create care centers. It supports clinicians in the field of imaging and pathology. In the imaging field, IBM introduced a technology called "Markation"; whose goal is to create the images and provide a suitable summary to support the radiologist more accurately [19]. Its motto is to reduce the burden of recording the patients' data and time in evaluating the radiologic images, furthermore, utilizing next-generation sequencing to identify and report the details of cases and provide the appropriate care to them [20].

Figure 8: Cognitive computing in healthcare.

Education

Cognitive computing has a greater footprint in the education field too. It is going to alter the education field. Already, a few changes have been made in this field to improve the education of students. It will alter how the universities, colleges, and schools have been working. Also, it provides study materials to the students for their easy learning purpose. Also, it can search the required data from the library much faster than from a digital library world. Furthermore, the cognitive system conducts tutorials for students, guiding them on how to learn and support the students to learn the subjects. Also, it makes the student select the course of their interest. So, it will act as a student counselor. It not only supports the students but also supports the teacher for their better service.

Chatbots

It is a program that can simulate human brain's thinking and processing ability in a contextual sense of expression and understanding. Here, a generic NLP approach has been utilized. NLP permits the program to collect the input from a human, clarify it, and make logical answers. Cognitive computing makes chatbots an appropriate grade in communication. Likewise, it estimates the users' needs depending on the previous communication.

Analysis by Sentiment

It has been found in the advancement in computing the recent times that the ability of science and technology in analyzing emotions through communication. It has been found that, in the advancement of science and technology it analyzes the emotion through communication. It has been considered a more complicated one for the machines. To tackle this complicated task, the machine should be trained with a set of individual interactions, then examine for its accuracy. It has been widely used in social media communications such as tweets and reviews [21]. The sentiment analysis in social media is gaining momentum as this can be leveraged to use marketing to the fullest potential. Figure 9 narrates the sentiment analysis by the cognitive computing system.

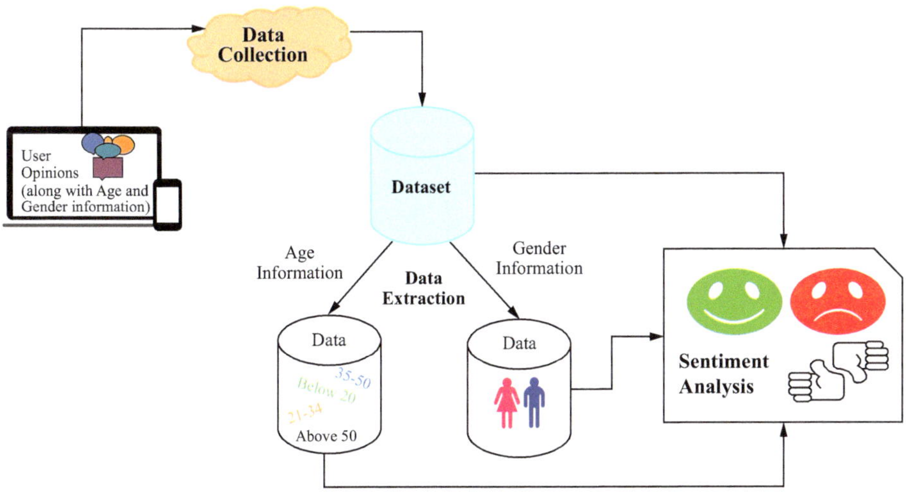

Figure 9: Cognitive computing in sentiment analysis.

Face Detection

The cognitive system is also supportive of face detection. It utilizes input information such as contours, texture, and eye color to discriminate an individual from others. If the face image is available, it can then be utilized to recognize the face from the still image or video content. In the previous method, this work was utilized in 2D image format. But now, the same set of operations has been performed on the 3D image, which results in greater accuracy.

Risk Management

In financial service, it necessitates the analyst to check over the current trends in the marketplace, previous data, and so on to evaluate the lack of security associated with the additional source of income. But it is not directly involved in huge data analysis. So, an intelligent algorithm is necessary to do this task. Cognitive computing performs this task which combines both behavioral data and trends in the marketplace. Afterwards, it can then be evaluated by the specialist for further predictions.

Fraud Detection

While considering finance, it tends to be another application. It is a kind of anomaly identification. Its motto is to recognize the transaction; its operation has been depicted in Figure 10. For this, there is a need for a program to check the previous history to know about the parameter to analyze the transaction details. Furthermore, the huge size of the dataset is required for classification. Various methodologies utilized here for classification are logistic regression, clustering, random forest, and so on. Data science as an individual entity of today's business and marketplace gained reputation as it could make meaningful predictions and reduce operation overheads through proper insights obtained from analyzing the financial data.

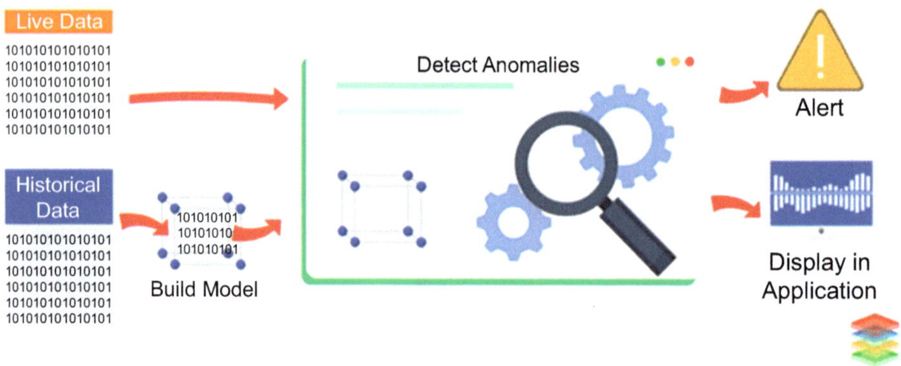

Figure 10: Real-time anomaly detection.
Source: https://www.xenonstack.com/blog/real-time-anomaly-detection [2019]

Top Cognitive Computing Companies

Cognitive computing utilizes AI to perform problem-solving skills like a human. It also decides to overcome the challenges faced in the business, industries, medical field, and so on [22]. Usage of cognitive computing in the education and medical field (healthcare) plays a key role in solving real-world problems in these areas. Upon considering the education, it needs a peer involvement that is 1:1 ratio in guiding the student. Furthermore, in the medical field, it supports the expert to take the health-related issues in a timely manner by analyzing the individuals' records in a short period of time. Some of the cognitive computing companies are given below for the reference:

Vantage Software

It has introduced a product called Coalesce, which is an IBM-based Watsons' Cog computing invention. Using this, the analyst can go through millions of documents and pages to know about the current market strategy, risk, and financial profile.

Welltok

It introduces a tool named Cafewell Concierge. It responds to individuals' inquiries in a few seconds with appropriate suggestions.

Lifelearn

It is a veterinary decision tool; it takes care of the animals with an immediate suggestion about the animals' health condition. It supports busy veterinary surgeons to save their precious time.

Wayblazer

This tool will support the travelers as a caretaker. Upon the request gathered from the travelers, cognitive computing will respond with different possible results. The request may be garnered in an "n"-dimensional array and offer variety of solutions for the needs.

Edgeup Sports

About millions of people love fantasy football. Huge among them invest a lot of time in reading the recently updated details. Some administrators utilize Edgeup sports, a cognitive computing technology that creates drafts and shares among their team. It is nothing but an advanced sports app; it allows raising queries related to fantasy football. It provides the optimal response, which supports the manager's intake a proper decision.

Bright Minded

Most recent fitness coaches utilize a paper and pen approach to create the fitness or diet chart for the customer to attain their goals. Bright-minded has a psychological app, which is known as TRAIN ME. It is created mainly for clinical purposes and keeps on tracking the customers' health conditions to identify which is good for them. Afterward, it authorizes the health-fitness chart to the customer. It allows the trainers to send the information and provides training.

Spark Cognition

Spark Cognition delivers enterprise-scale AI solutions that enable organizations to optimize processes, predict future outcomes, and prevent cyberattacks. It utilizes sensors to identify the failure that may happen in the coming days and reduces such happenings by changing the operators.

Deep armor: The malware issues have to be reduced by ML approaches. The Deep Armor improves the performance and reduces the forthcoming issues.

Deep NLP: It utilizes a superior ML algorithm and combines the state-of-art approach to earn the solution for the issues.

Darwin: It utilizes the ML algorithm to create a prototype environment for data scientists in developing ML applications. It is said to compare against the other previous data science replica.

Numenta

It is developed based on the neocortex theory. It is also an ML technology. It identifies any deviation in the server's functionality, categorization of language, and behavior of humans with followup data.

Expert System

It provides the knowledge to utilize the constructed data by fusing both technology and skill. It supports both the occupation and government by making decisions. Cogito is nothing but cognitive software that provides individuals to examine and know about meaningful decisions. Their bases depend on the discernment of text. For better analysis, cognitive software combines with an ML algorithm.

Customer Matrix

By diffusing cognitive computing and revenue influence on the company, IT investment gets multiplied. It provides modified actions to every relationship manager to be aware of the technology. Thus, everyone can receive high revenues from that.

Challenges

Cognitive computing has enormous applications in various fields. Even though there exists a limited analysis of risk in unstructured data; it incorporates culture, people, political environments, and so on. For example, if the governance is changed in a country, a complete update needs to be considered. In such cases, human involvement is necessary for risk analysis. Furthermore, cognitive computing requires extra data for training. The challenging task involved here is the adaption of the training data. Since, the training of the cognitive system has been done gradually. Furthermore, the complex process of utilizing the cognitive system worsens.

Another thing to be considered is that cognitive computing is regarded as a most efficient assistant like AI and intelligence automation. It improves the thinking of humans, but if one wants to take any decision means, it should depend on individuals' opinions. Another challenge to consider here is privacy. Accessing data is unsafe in the case of any organization in maintaining any confidential information. So, an appropriate criterion has to be taken into account to protect the data. Data privacy and data isolation is a primitive step that any organization need to adapt in the data-driven world. In all cases, human involvement is necessary together with cognitive computing. So, cognitive computing now enters into the automation process. It sets a target for cognitive computing to reach the extent of humans. But now there exist certain limitations that it should not adopt the rapid happens in the outside world; also, to meet the demands of the individual. The complexity arises as there is an increase in the count of data.

Globally, experts believe that the growth of cognitive computing plays an important role in enterprises functioning. Research says that the current US$4.5 billion value of money of AI gets double in the upcoming year. Furthermore, it occupies a greater place in the market. Also, Microsoft, Apple, Google, Amazon, IB, and so on have been involved in creating the forthcoming product. Moreover, the Government of India identifies the advantage of ML to examine student dropouts. ML did with the support of Microsoft's cloud-based analysis and Azure ML. Andhra Pradesh government introduced the project to collect and analyze the status of the students by collecting all the relevant data. The data contains the details such as teachers, school, and students' family backgrounds. It supports the government to allocate the resources legally [23].

Conclusion

As a whole, cognitive computing has a vast innovation and continual improvement in AI and the adapted big industries. Also, it provides the solution to meet human demands: react, concentrate, and act as an individual to attain an optimal interaction of human and machine. In this chapter, the various applications of cognitive computing systems and their challenges involved have explained. We might think that soon digital systems will engage a lot in the cognitive systems. Cognitive computing will become a crucial one in the digital rationalist outlook. Any organization that wishes to grow continuously concerning speed and scale of the activities must introduce the cognitive computing facilities themselves. It will act as the next step for any organization that follows any traditional activities which run the human involvement to reach the milestones of growth in a quicker time.

References

[1] B. Krzanich, The intelligence revolution – Intel's AI commitments to deliver a better world. Web-Link Editorial, Intel, 2016. Newsroomhttps://newsroom. http://intel.com/editorials/krzanich-ai-day/
[2] W. Raghupathi, "Big data analytics in healthcare: Promise and potential," Health Information Science and Systems, vol. 2, no. 1, pp. 1–10, 2014.
[3] J. S. Hurwitz, M. Kaufman, and A. Bowles, Cognitive Computing and Big Data Analytics, John Wiley & Sons, 2015. ISBN: 978-1-118-89662-4
[4] D. S. Modha, R. Ananthanarayanan, S. K. Esser, A. Ndirango, R. Singh, and A. J. Sherbondy, "Cognitive computing," Communications of the ACM, vol. 54, no. 8, pp. 62–71, 2011.
[5] V. Raghavan, V. Gudivada, V. Govindaraju, and R. R. Rao, Cognitive Computing: Theory and Applications, Handbook of statistics, Elsevier, Vol. 35, 2016.
[6] J. E. Kelly, Computing, Cognition and the Future of Knowing [White Paper], IBM Research, pp. 1–7, 2015.

[7] Y. Wang, "Keynote: on cognitive informatics," in Proceedings of the 1st IEEE International Conference on Cognitive Informatics (ICCI2002), Calgary, Canada. IEEE CS Press, 2002, August, pp. 34–42.

[8] Y. Wang, "On cognitive informatics," Brain and Mind, vol. 4, no. 2, pp. 151–167, 2003.

[9] Y. Wang, "Keynote: Cognitive informatics – Towards the future generation computers that think and feel," in Proceedings of the 5th IEEE International Conference on Cognitive Informatics (ICCI2006). 2006, pp. 3–7.

[10] Y. Wang, "The theoretical framework of cognitive informatics," International Journal of Cognitive Informatics and Natural Intelligence, vol. 1, no. 1, pp. 1–27, 2007.

[11] Y. Wang, "On abstract intelligence: Toward a unifying theory of natural, artificial, machinable, and computational intelligence," International Journal of Software Science and Computational Intelligence, vol. 1, no. 1, pp. 1–17, 2009.

[12] Y. Wang, "On cognitive computing," International Journal of Software Science and Computational Intelligence, vol. 1, no. 3, pp. 1–15, 2009.

[13] Y. Wang, "Cognitive robots: A reference model towards intelligent authentication," IEEE Robotics and Automation, vol. 17, no. 4, pp. 54–62, 2010.

[14] Y. Wang, "Towards the Synergy of cognitive informatics, neural informatics, brain informatics, and cognitive computing," International Journal of Cognitive Informatics and Natural Intelligence, vol. 5, no. 1, pp. 75–93, 2011.

[15] P. Adhi, B. Mutell, A. Davis, T. Burns, and S. Lal, Future of Retail Operations: Winning in a Digital Era, McKinsey & Company, 2020.

[16] T. Roberts, Power of cognitive banking: Excited to announce our next customer with Personetics – Israel Discount Bank, 2018.

[17] N. Sita Ramudu and M. Nalini, "A user-centric machine learning for cyber security Operations Centre," Test Engineering and Management, vol. 82, p. 87, 2020.

[18] https://www.techfunnel.com/information-technology/applications-of-cognitive-computing-technology/.

[19] A. D. Norden, K. Rhee, A. Urman, I. Dankwa-Mullan, and F. Suarez, "Realizing the promise of cognitive computing in cancer care," User in a New Era, JCO Clinical Care Informatics, vol. 2, no. 1–6, 2022.

[20] N. M. Patel, J. M. Snell, V. V. Michelini, A. P. Hoyle, S. Balu, M. C. Hayward, J. S. Parker, A. H. Salazar, D. A. Eberhard, J. Xu, P. McNeillie, T. Koyama, C. S. Huettner, R. Norel, F. Utro, K. Rhrissorrakrai, A. Royyuru, E. Bilal, H. S. Earp, and W. Y. Kim, "Enhancing next-generation sequencing-guided cancer care through cognitive computing," Oncologist, vol. 23, no. 179–185, 2018.

[21] S. Kumar, M. Gahalawat, P. P. Roy, D. P. Dogra, and B. Kim, "Exploring impact of age and gender on sentiment analysis using machine learning," Electronics, vol. 9, no. 2, 2020.

[22] https://www.hitechnectar.com/blogs/cognitive-computing-products/.

[23] https://analyticsindiamag.com/building-cognitive-computing-capabilities/.

R. Ranjana, Subha T, Varsha, Jothishree, Pravin Kumar

Applications and Implications of Artificial Intelligence and Deep Learning in Computer Vision

Abstract: Recent advancements in machine learning have enhanced the features several applications. One such field is the computer vision (CV). The recent times have seen the deployment of robots in many industries and household. Vision for robots or robotic vision is also an emerging field of research.

CV can be characterized as a set of errands that incorporate styles for obtaining, deciding, taking apart and grasping computerized pictures, and parentage of high-layered information from this present reality to deliver mathematical or authentic information, for example, in the types of liberations. Understanding in this context-ture implies the transformation of visual pictures (the contribution of the retina) into portrayals of the world that appear to be legit to concentrate on processes and can inspire relevant activity. This picture understanding should be visible as the unraveling of significant data from picture information utilizing models developed with the guide of figure, medications, insights, and learning recommendation.

The logical discipline of CV is worried about the speculation behind counterfeit frameworks that wine tool data from pictures. The picture information can take innumerable structures, looking like live video or video recording arrangements, sees from different cameras, multilayered information from a 3D scanner, or clinical examining gadget. The innovative discipline of CV looks to apply its recommendations and models to the development of vision frameworks.

Deep learning and other artificial intelligence techniques can be used to enhance CV. This chapter aims to bring out the various methods and algorithms that are used in enabling a computer to see.

Keywords: Computer vision, Deep learning, frameworks, machine learning, vision based systems

R. Ranjana, Department of Information Technology, Sri Sairam Engineering College, Chennai, e-mail: ranjana.it@sairam.edu.in
Subha T, Department of Computational Intelligence, SRM Institute of Science and Technology, Chennai, e-mail: subharajan@gmail.com
Varsha, Jothishree, Pravin Kumar, Sri Sairam Engineering College

https://doi.org/10.1515/9783110750584-003

Background and Introduction to Computer Vision

Computer vision (CV) is the fastest developing area in technology. CV is a subfield of artificial intelligence (AI) and machine learning. When we say machine learning, deep learning is also a part of it. Humans are capable of understanding the problems and describing them in the form of an image.

If a two-dimensional image is given to the CV system, it must recognize or detect. The object features must provide the complete description of the object. When we say CV, the first thing that strikes us is image processing. Image processing is the most important field in CV. CV is the transformation of information from a digicam into an information that can be communicated as "the digicam is mounted in a car" or "laser variety finder shows an item is 1 meter away" both a choice or a brand-new representation. All such modifications are executed to achieve a few specific goals. The entered information may also consist of a few contextual information [1].

Image processing is a technique of applying some transformation to the raw image. The main aim of CV is to improve image quality and explain about it. A CV system can prepare an image input and give us a clear briefing about that image. For example, noise reduction, haze and blur removal, rotational operation, image processing, and capturing can be performed at pixel level. CV can solve more problems such as object detection and behavioral tracking. If we search an image in Google by scanning, the reason and technology behind that is deep learning with CV. Most pc scientists and realistic programmers are aware about a few aspects of computer vision's role; however, few humans are aware about all the methods wherein pc vision is used. For example, maximum humans are fairly aware about its use in surveillance, and many additionally realize that it is miles more and more getting used for pictures and video on the Web [2].

Application of Computer Vision

Humans cannot understand the image and scene, and give a clear briefing of that image. In that case, CV has a large range of applications [3]. As we know, CV depends on AI and machine learning.

Retail and Shopping

In recent years, the software of laptop imaginative and prescient with inside the retail area has been the maximum sizeable trend. The possibilities in the retail arena range from clothing to food and household supplies. Shopping on the Internet is simple. Customers can browse for things they want to buy in a visually appealing and simple manner. Custom machine learning technology can help clients have a better experience.

Behavioral Tracking

Face recognition and object detection are both possible using algorithms. It is CV's primary goal and scope. Gender, age, height, weight, color, and other human traits are determined using face recognition algorithms. CV algorithms combined with store cameras are used by brick-and-mortar merchants to analyze their consumers and how they act.

Healthcare

There are a variety of CV applications in the healthcare field. Medical image analysis is the most well-known example. To discover malignancies and neurological disease, images from MRI, CT scans, and X-ray are evaluated. It all comes down to an image analysis technique that extracts data from photos in order to identify the correct problem or disease. This type of application is currently in use by a huge number of people employing sensing and detecting equipment to quantify the volume of blood lost in an artery and evaluate blood flow in a human body.

Autonomous Vehicles

Detecting and finding objects is the most intriguing challenge in CV. This requires the ability to locate objects as well as the ability to classify and identify them. This technology is used in conjunction with data processing from sources such as sensors to allow objects to "see."

Need for Computer Vision

Problem resolution is an essential objective of CV. Self-driving automobiles can sense their surroundings, thanks to CV. The fastest growing technology for sensing your surroundings is vision. Perhaps this is why complex life forms evolved eyes for sensing purposes. The simplest and least expensive method of capturing light and "seeing" the environment is to use a camera. The cost of Sophisticated techniques for sensing the environment are no longer the bottleneck but the need to do the same must be justified. One of the hottest topics in robot sensing is CV. Unless we want to make dumb, nonautonomous robots, the only way to get a good feel of your surroundings is to use CV. With this technology, we also use sensor fusion to sense. They incorporate omnidirectional vision for robots and independent vehicles, infinitesimal relapse issues, and worldwide climate and site demonstration. An innocent programming of convolutional orders to a planar convexity of the round sign is bound to fall flat, because of the reality that the space-different bends added with the attendant of operating this kind of protrusion will make translational weight partaking deficient [3].

Aims of a Computer Vision

CV can be used to detect, localize, segment, and recognize certain objects.

The process of mapping a three-dimensional model to a scene that is used to navigate a robot using image sensing.

Following an object through a series of images

The content of digital photos can be searched for (content-based image retrieval)

Statistical learning, pattern recognition, graph theory, picture classification, image processing, and other topics can help attain these aims.

CV has proved its ability to revolutionize practically every area with the various applications that have been employed around the world. Retail, healthcare, automotive, engineering, and manufacturing, among other industries, have recognized the value of CV and AI in improving their processes and customer experiences. According to a forecast estimate, CV software would expand to USD 26 billion by 2025. Hand gestures are a shape of nonverbal communique that may be utilized in numerous fields which includes communication among deaf–mute people, robotic control, human–pc interaction (HCI), domestic automation, and scientific applications [4].

Implications of AI Techniques in Computer Vision

This pc imaginative and prescient program is utilized by means of synthetic intelligence and mimics how people see and understand, and provide an explanation for snapshots – a vicinity of synthetic intelligence wherein a laptop is taught to understand the visible world. Using virtual snapshots from cameras, movies, and deep studying models, machines can become well aware of and discover objects, then react to what they observe. This era primarily based on laptops and the imaging available because of it, it is rather easy to correctly acquire statistics from the background environment of the pictures or videos. Things and phenomena within the global goal have been analyzed, judged, and decided [5]. The reason for photo evaluation within the late 1950s and early 1960s became to emulate human imaginative and prescient structures and ask computer systems what they saw. Computer generation became utilized to hastily and appropriately gather crucial records from the surroundings so as to analyze synthetic intelligence, primarily based totally on laptop imaginative and prescient imaging, with snapshots and videos. Computer imaginative and prescient is the capacity to see. In the goal, world matters and occurrences had been analyzed, judged, and decided. The middle ideas and ideas of laptop imaginative and prescient had been investigated for the use of organic imaginative and prescient and bodily models. The researchers checked out the contents of evolutionary calculations, sample recognition, organic imaginative and prescient, and slab floor fault detection. A laptop's imaginative and prescient machine became created with exquisite adaptability, precision, and pace of response. The findings discovered that evolutionary calculating techniques got here in plenty of sizes and styles and had been pretty adaptable. This AI generates three-dimensional high-decision reconstructions of human beings from two-dimensional images! It wishes for an unmarried photograph of you to generate a three-dimensional avatar that appears much like you, even from the back [6].

How Does AI's Computer Vision Work?

- To produce the findings, AI CV follows these three steps:

Image acquisition is the process of converting analogue data into a computer-readable format, that is, a series of zeros and ones. Datasets are prepared:
- Using digital cameras, webcams, and a variety of other instruments.
- Image processing involves extracting geometric elements from a picture and the usage of contemporary-day implemented arithmetic strategies inclusive of segmentation, classification, part detection, and characteristic identity and matching.
- Image evaluation and comprehension: For three-dimensional scene mapping, item tracking, and recognition, advanced-degree algorithms are used to adopt deep evaluation of the data.

AI assists CV in accomplishing the following goals:

Estimation of Human Pose

CV technology can detect and evaluate human posture with the help of AI. It is primarily used in fitness coach apps to capture the user's movements while exercising, calculating the accuracy of their body posture, and recognizing and correcting their faults.

An Examination of Video

Video analysis provides you with additional use cases for gaining a better knowledge of a scenario. It is used to monitor objects, classify behaviors, and used more in security camera systems. This has all been made possible thanks to CV.

Automobiles that Drive Themselves

AI-based CV can perceive the environment and recognize various items on the road, such as pedestrians and traffic lights. It also aids in determining the distance between the car and other vehicles.

Recognizing People by Their Faces

The technology allows a device to recognize a person's face and authenticate their identity. It then searches the stored database for a match after detecting the face. Apart from unlocking the device, facial recognition also assists social media apps in suggesting users to tag in a picture or video.

Virtual and Augmented Reality

In gadgets inclusive of smartphones and tablets, laptop imaginative and prescient additionally aids in protecting and embedding digital matters on real-global imagery.

Text Recognition

Your image recognition system uses AI to identify messages from photos. It might be quite useful in converting old books and papers to digital format. Text recognition benefits every industry, from retail to logistics and marketing. Furthermore, industry behemoths are significantly investing in this technology in order to gain future benefits.

Deep Learning and Computer Vision

In the field of CV, deep learning has achieved remarkable progress in a short time. Face recognition systems, self-driving cars, and other inclusion of deep learning in CV are only a few examples. CNN is covered in the deep learning application since it is the most significant aspect of deep learning with CV. Deep studying has fueled top notch strides in a number of pc imaginative and prescient problems, consisting of item detection [7].

Convolutional Neural Networks

Convolutional neural networks are like ordinary phase networks in that they are neurons with learnable loads and inclinations, as clarified in the past section. Every neuron takes few data sources, does a dab item, and afterward, whenever required, plays out a nonlinearity. The general employer determines a solitary differentiable rating work, from crude photograph pixels in the direction of one aspect to elegance ratings at the other. Applying the convolutional activity to picture information is not new or selectable to convolutional neural organizations; it is a typical PC vision approach. PC vision experts used to fabricate channels by hand, which they then, at that point, applied to an image to get the ideal outcome. Somehow or another, the element guides or yields from applying the channel then, at that point helps picture investigation. The first computational fashions primarily based totally on those neighborhood connectivities among neurons and on hierarchically prepared ameliorations of the picture are discovered in neocognitron [8].

Example

Assume we have a contribution of shape 32 × 32 × 3.

There is a mix of convolution and pooling layers toward the start, a couple of completely associated layers toward the end, and lastly a softmax classifier to characterize the contribution to different classifications. There is a great deal of hyperparameters in this organization which we need to determine.

By and large, we take the arrangement of hyperparameters which have been utilized in demonstrated examination and they wind up progressing nicely. As found in the above model, the stature and width of the info shrivels as we go further into the organization (from 32 × 32 to 5 × 5) and the quantity of channels increases (from 3 to 10).

Semantic Segmentation

Semantic picture division is the assignment of grouping every pixel in a picture from a predefined set of classes. To perform semantic division, a more elevated level comprehension of the picture is required. The calculation should sort out the items present and furthermore the pixels which relate to the object. It consists of making jumping confines around the articles containing a picture and characterizing every single one of them. A large portion of the item location models use anchor boxes and propositions to recognize bouncing boxes around objects.

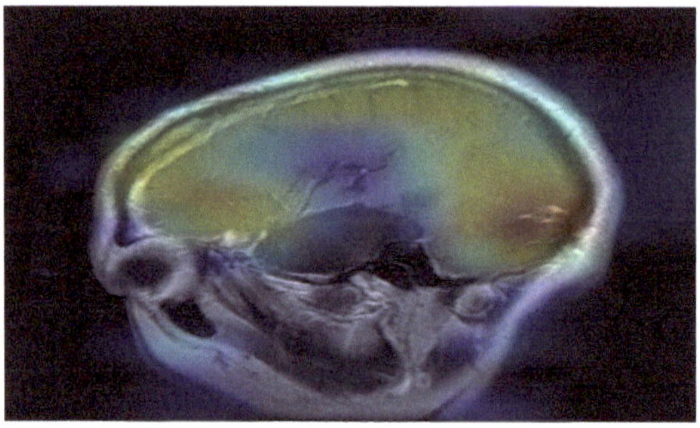

Figure 1: An image representing the disease segmentation which is given as example for semantic segmentation.

Recurrent Neural Networks

Recurrent neural network (RNN) recollects the beyond and its selections are impacted with the aid of using what it has won from the beyond. Note: Basic feed ahead networks "recall" matters as well, but they consider matters they were given the hold of for the duration of preparing. For instance, a photograph classifier realizes what a "1" looks like for the duration of training and later on makes use of that data to organize matters underway. It is important for the organization. RNNs can take at least one info vector and produce at least one yield vector and the output(s) are affected not simply by loads applied on inputs like a customary neural network (NN), yet in addition with a "covered up" state vector addressing the setting dependent on earlier input(s)/output(s). Thus, a similar info could deliver an alternate yield contingent upon past contributions to the series. The diagrammatic representation of the working of CNN is explained in Figure 2.

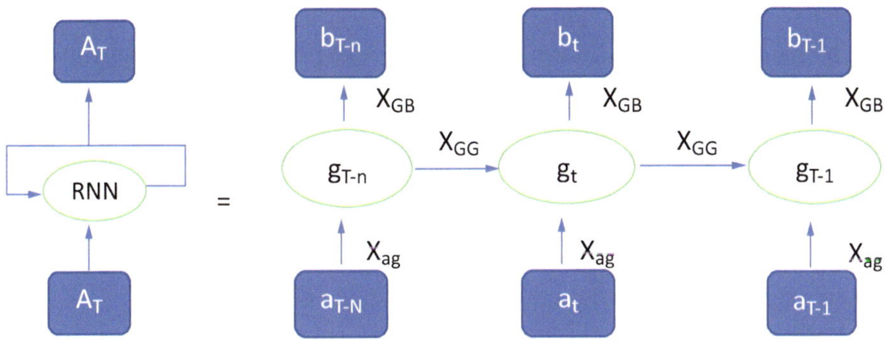

Figure 2: Working Steps in RNN.

Methods Used in Making Computer "See"

CV is a scientific field of engineering science that deals with the ideas and techniques which are developed to assist the system to recognize pictures such as photographs and videos. We humans have the ability to see through our natural vision, recall and summarize a protracted video to a brief one, but machines lack of those capabilities, hence the concept of CV is introduced here to see. American sign language recognition uses input as 3D arms motions to capture and process it using CV which uses 3D parameters extraction and assumptions to deliver the output [9].

In the contemporary world, CV algorithms work on the principle of pattern recognition like our brain which compiles the source image with eyes and decodes it by separating the objects into patterns. By making a computer see, it can be fruitful for the user; for example, if we want to locate the fruits on trees, CV can be used to identify the variety of fruits the user needs [10]. In addition to that, machines observe the pixels obtained from the images which have a unique set of values of colors.

For example, if we send 1,000,000 pictures of butterflies, the system will analyze the patterns present in butterflies and find the match which looks similar to a butterfly. As a result of this process, the computer can identify a particular image as a butterfly when it is given as an input. Deep learning in CV has created quick progress over a brief amount. Deep learning is employed in numerous ways in CV, such as face recognition gadgets, object detection systems, and self-driving cars. Convolutional neural networks, conjointly referred to as convnets, is a kind of deep-learning model universally employed in CV applications and understanding its elements, layers like convolutional layer, pooling layers, and totally connected layers.

As a child, we have been taught to classify things by pattern recognition. As our brains have a tendency to develop learning, analyzing and detecting a lot of living and nonliving beings around us, it registers the assorted options based on some of the main characteristics like structure, geometry, color, and size of the objects. If we tend to see a picture multiple times possessing those options, we are ready to predict what actually the object is and finally that is one thing we have learned with expertise. Now here is the actual implementation of our concept in CV: we have to construct those parameters to computers in which our brain helps us to identify objects wherever we tend to try and determine the thing gift within the image. If a picture is taken, first computers ought to be ready to extract its options so supported those options, predict on what that object is.

Deep Learning Algorithm Used in Edge Detection

Edge detection is a picture processing method that is used to become aware of factors in a virtual picture with discontinuities, sharp modifications within the brightness of images. The picture brightness that varies sharply are called edges of the picture. The visualization entities from sparse backgrounds with the aid of incorporating edge detection into the prevailing pipeline [11].

The simple steps in picture processing are sample popularity and laptop vision. Edges occur between two regions. The feature is the extraction of edges of an image. Edge detection has the main feature for analyzing the image [12]. When we use this method on very high-decision virtual images, convolution strategies come to the rescue.

Object Detection

Object detection includes the aggregate of item category and localization inside snap shots or videos. CV lets in pc structures to "see," detects the items in surrounding regions, and item detection simplifies the process. CV is a category of AI.

Pattern Recognition

Pattern recognition is the technique with the aid of which a device acknowledges the facts. The type is primarily based totally at the facts of histories, statistical information, and the reminiscence of machines. Machine getting to know algorithms is the method utilized in sample popularity.

Use Cases for Computer Vision

CV is an AI technique. CV is the ability for computers to recognize and label images. CV is utilized in autonomous car testing, agricultural, livestock, and convenience store health monitoring. In our investigation, we discovered that computers are capable of recognizing photos. We chose to investigate how the world's leading firms employ CV, as well as what new technology and media might emerge as a result. Google, Amazon, Facebook, and Microsoft are among the leading technology corporations investing billions of dollars in CV research and product development. In our CV research, we discovered that several of the applications of CV include:

Retail and Security in Retail

Amazon just introduced the Amazon Go store, which allows customers to pay for their items without having to stand in line at the checkout counter. This Go shop is equipped with cameras that are mostly used for CV. Only Amazon employees are permitted to shop. The technology that underpins this Amazon Go store is called Just Walk Out.

Automotive

According to the World Health Organization, 1.25 million people die each year as a result of road accidents. Vulnerable road users use nearly half of the roads. This trend is expected to become the sixth biggest cause of mortality in the next several years. Human mistake and inattention are a common motif in traffic accidents, according to the study.

Agriculture

Farms are starting to embrace CV technologies to help them run more efficiently. These technologies are designed to assist farmers in adopting more efficient growing practices, resulting in higher yields and, ultimately, higher profits.

Banking

CV technologies can also be used in the banking industry. Fraud detection and natural language processing are two areas where banking has been involved.

Deep Learning Algorithm for Image Segmentation

The key material in photograph process and portable computer creative and discerning is photograph segmentation. Varied algorithms for photograph segmentation were advanced in literature. The programs are increased reality, scientific photograph analysis, golem perception, and evaluating the images. Layout and exposition in addition create it eminently applicable as a totally distinctive reference to the essential methods and newest studies literature in laptop vision [13]. We glance at the similarities, strengths, and contests of the deep gaining information of fashions, the most extensively used datasets, performances of stories on this area. There are a lot of paintings geared toward photograph segmentation improvement processes and the usage of deep gaining information of fashions [14].

Deep Learning Algorithms for Object Identification

In phrases, we will say that the intention of item detection is to come to a decision whereby gadgets are placed for the given icon is named as item localization. There is a quick extrude within the discipline of laptop computer ingenious and discerning in item detection. Object detection is disquieted within the kind of gadgets. The category wherein each item belongs is referred to as item type. This technique detects the item and locates its times for photos and flicks in virtual form. Deep learning may be a moneyed own circle of relatives of methods, encompassing NNs, hierarchical probabilistic models, and quite a few unattended and supervised characteristic learning algorithms [15]. The capability of detection and pursuit targets will play a major role in mobile golem navigation systems. Visual tracking systems could manage the direction and speed of motion of a robot to stay at the target in its field of vision either by moving the robot itself or the vision sensors [16].

Role of Computer Vision in Augmented Reality and Virtual Reality

Augmented reality (AR) is a huge benefit to all the industries and virtual reality (VR) is created by gamers for their gaming community. Even though both serve different audiences, we should look upon these technologies in our near future where it paves the way for our modern society. VR immerses someone into a virtual world stimulating their gospel through senses. For assembling the components in an industry, the machine needs to visualize the materials in order to attach it. Hence, CV plays a preliminary role in deciding the construction of the machine to interpret the forthcoming sequence with the help of AR [17]. This stimulation is achieved through a supply of content and hardware like headsets, treadmills, gloves, and so on. CV aids video games with strong vision capabilities like simultaneous localization and mapping, structure from motion, user body chase, and gaze chase. CV assists humans in vast areas including VR and AR. One such application is the registration for the outdoor events using CV by tracking down the real-world original positions through a horizon silhouette [18].

Using cameras and sensors, these functions facilitate VR systems analyzing the user's atmosphere and finding the headset's location. So, CV and video games work along to create merchandise that is a lot refined and user-responsive.

Augmented Reality

AR has the potential to instill awe in the US by joining the physical and real worlds. In reality, CV-based AR overlays representational processes or audio onto the present real-world scenery. And it all begins with CV. CV for increased reality allows computers to get, process, analyze, and perceive digital videos and pictures. By gazing at the associate object and its look, location, and also the settings, it identifies what the item is. Instagram acknowledges your friends by exposure tags, the way you will be able to get yourself a flower crown on Snapchat.

For example, take Snapchat filters. After you check out photos, you see your face, whereas CV sees information. Machine learning, a technique of knowledge analysis, incorporated in associate AR app soothes the pain of object detection. CV, in turn, maps facial expression from photos or videos. It reads the structural geometry of your face, captures facial landmarks and conjointly takes into consideration the universal truths concerning the face like uneven options, mimic facial muscles, and so on. To remove a predefined AR content (be it a cervid face or hearts round the head), you must get an associate correct face scan. During this case, CV allows AR image process, optical pursuit, and scene reconstruction, which is significant for any immersive app. Next, a CV-based AR system scans your exposure with sensors to feature the time period visual effects to your face. It all brings up the combo of the physical world and AR information. In fact, CV goes higher than and on the far side operating closely with AR and VR to produce the users with subtle interactive content.

Virtual Reality

A VR tool will sight all of the frame actions of a user through high-velocity rendering; it is going to produce the phantasm of complete worlds. This immersive enjoyment can be an area for learning, undertaking experiments in an incredibly secure place, or possibly producing emotions. To create the phantasm of reality, it is important to appear inward and outward at the equal time. CV imaginative and prescient makes use of outside cameras or sensors to map the placing, but as eye-monitoring answers and gyroscopes to place the user. VR requires faster computations to place the person in the immersive environment based on his inputs. Any lag or put off in rendering is perceived as a flaw in the device and makes the person much less willing to examine the digital placing as real. A widespread benefit of this method of VR rendering is that the cost-effectiveness considering that maximum smartphone-kind cameras are extraordinarily reasonable. The stay feed from the digital digicam is analyzed through the set of rules that detects the method precise points, like corners, move. This will be nearly like what we have a tend to, as people do when we check distances and take a look out to keep away from objects. Through know-how fusion techniques, the CV generates a three-dimensional version of the placing, which is probably wont to provide an accomplice diploma accurate VR viewpoint.

Challenges in Building Vision for Computers

CV is combined with many techniques for interpreting resume image data. Computers cannot interpret these images in any way since image processing has been made practical for a very long time; for example, in order to provide semi-intelligent image processing functions, a camera which uses gamma analysis is used to check whether a certain region is infringed, but not more than that. The use of artificial neural networks has increased the potential of machines to unprecedented levels. Now, the computer can understand the actual happenings around its environment or a specific scene and it uses the application of noise cancellation to update more accurate information for the user removing unwanted sounds which is obvious. CV assists humans in low-power solutions to detect objects in images to get rid of power-consuming batteries that machines depends on [19]. A child running amidst people or the car coming ahead of us is highly considered rather than birds flying in the sky or the forthcoming gasoline station. There are various techniques and challenges in building vision for computers. It is essential that we overcome those obstacles to avoid malfunctioning of the machines [20]. The fact that CV in the context of a car can distinguish between trucks and people, or buildings and trees does not preclude that the perception of us (humans) competes. Skills are just a preconception to more complex techniques, but they are not without challenges. The following are some of the problems faced by CV:

Challenge one: Automatic sensors and multimodal data
Challenge two: Collect information representing training data
Challenge three: Image and object recognition
Challenge four: Semantic segmentation task of instances
Challenge five: Stereo vision and multiple camera vision

Challenge six: Tracking of Objects and images.
Challenge seven: Three-dimensional scene analysis with AI.

In this contemporary world surrounded by developed technology, as long as we meet the challenge, there are unlimited possibilities. Given the humongous growth in CV abilities, it is a piece of cake to see new areas of applications have emerged. Self-driving automatic cars are one of the toughest that is more challenging to make it work but also the most enlightening fields in the application of modern CV technology yet they have their own disabilities too. Collect input files for specific tasks to create deep learning experiment with your favorite frameworks reminiscent of Pytorch, Tensorflow, and so on. Prepare the environment by choosing the right hardware like Intel, ARM, NVIDIA, and platforms such as Linux, Windows, Docker, Kubernetes, and enhanced learning models. Cloud, that is, device cloud, manages the deployments in hundreds of edge gadgets. Organization and deployment improvements are distributed across the entire edge endpoint cluster. All endpoint indicators and analysis of data comply with the best practices of knowledge security and data protection, as time goes by high-risk developments that may be connected use this method, especially noticing the design time, knowledge in domain, and climatic design difficulties. Real scenes using CV's hardware function for operation and for visual information, cameras are used. The only way to overcome the disabilities is to inculcate AI gadgets for edge intelligence. CV is mainly focused for rapid development of machine learning, the number of AI use cases working at personal or higher levels has grown exponentially. AI vision includes technology used in machine vision business to solve a series of continuous problems, deep learning, victimization, and CV; however, the best creative potential of abstract thinking in AI requires clean computing power, especially for periodic data-intensive applications. In addition, AI resolutions are provided in some of the famous tech giant's cloud services (e.g., Google, Amazon, and Microsoft) to take advantage of minimized and scalable computing resources.

Conclusion

In conclusion, CV is not a cup of tea and it requires a large attempt to begin digging deep in that universe. In building a vision for computer, we may face lot of challenges as mentioned above. Our sole ideology for our project is to make a computer "see" through and detect objects through CV using deep learning strategies. By aggregating the data above, we develop a robust prototype which can be used to predict objects placed in front of the computer.

Case Study

Problem:
In image procect on a complex background in critical temperature and weather conditions.

Solution:
The proposed solution for the framework includes a chain of picture processing and imaginative and prescient strategies. Hence, choice of picture processing and laptop imaginative and prescient strategies is a vital venture within the proposed framework. So, we are suggesting that the subsequent strategies that need to be covered within the processing primarily based totally on exclusive environmental conditions.

References

[1] Ahonen et al., "Face recognition with local binary patterns," in The European Conference on Computer Vision. Springer, Berlin, Heidelberg, 2004.
[2] Agrawal et al., "CenSurE: Center surround extremas for realtime feature detection and matching," in The European Conference on Computer Vision, 2008.
[3] S. C. Taco et al., Important Computer Vision Research Papers of 2018 Spherical CNNs.
[4] A. Al-Naji et al., "Hand gesture recognition based on computer vision: A review of techniques," Munir Oudahl Journal of Imaging, vol. 6, no. 8, p. 73, 2020.
[5] S. S. Rautaray and A. Agrawal, "Vision based hand gesture recognition for human computer interaction: A survey," Artificial Intelligence Review, vol. 43, no. 1, pp. 1–54, 2015.
[6] S. Saito et al., "Pifuhd: Multi-level pixel-aligned implicit function for high-resolution 3d human digitization," Proceedings of the IEEE/CVF Conference on Computer Vision and Pattern Recognition, pp 84–93, 2020.
[7] Ouyang, Wanli, et al. "DeepID-Net: Object detection with deformable part based convolutional neural networks." *IEEE Transactions on Pattern Analysis and Machine Intelligence* 39.7 (2016): 1320–1334.
[8] K. Fukushima, "Neocognitron: A self-organizing neural network model for a mechanism of pattern recognition unaffected by shift in position," Biological Cybernetics, vol. 36, no. 4, pp. 193–202, 1980. 39, no. 7, pp. 1320–1334, 2017.
[9] C. Vogler and D. Metaxas, "Adapting hidden Markov models for ASL recognition by using three-dimensional computer vision methods," in IEEE International Conference on Systems, Man & Cybernetics, Computational Cybernetics and Simulation, 1997, vol. 1, pp. 156–161.
[10] A. R. Jimenez and R. Ceres, "A survey of computer vision methods for locating fruit on trees," Jose L Pons Transactions of the ASAE, vol. 43, no. 6, pp. 1911, 2000.
[11] F. Wang, ETRA '20 Short Papers: ACM Symposium on Eye Tracking Research and Applications, June 2020.
[12] R. Szeliski, International Journal of Computer Science & Information Technology (IJCSIT), vol. 3, no. 6, Dec 2011.

[13] S.-Y. Chen et al., "DeepFaceDrawing: Deep generation of face images from sketches," ACM Transactions on Graphics, vol. 39, no. 4, p. 72, 1–72:16, 2020. BN:978-1-84882-934-3.

[15] A. Voulodimos et al., Deep Learning for Computer Vision: A Brief Review, Department of Informatics, Technological Educational Institute of Athens, 12210 Athens, Greece.

[16] M. Karam et al., "Video-based object detection and tracking system for weight sensitive UAVs," Multimedia Tools and Applications, vol. 78, pp. 9149–9167, 2019.

[17] R. Sharma and J. Molineros "Role of computer vision in augmented virtual reality," in Proceedings of SPIE 2409, Stereoscopic Displays and Virtual Reality Systems II, 30/03/1995.

[18] R. Behringer, "Registration for outdoor augmented reality applications using computer vision techniques and hybrid sensors," Proceedings of IEEE Virtual Reality, 1999, pp. 244–251.

[19] S. Alyamkin et al., "Low-power computer vision: Status, challenges, and opportunities," IEEE Journal on Emerging and Selected Topics in Circuits and Systems, vol. 9, no. 2, pp. 411–421, June 2019.

[20] S. Xu et al., "Computer vision techniques in construction: A critical review," Archives of Computational Methods in Engineering, vol. 28, pp. 3383–3397, 2021.

R. Regan, P. Rajaram, K. Senthilkumar, N. Khadirkumar

A Study of Voice Recognition System Using Deep Learning Techniques

Abstract: As there has been a tremendous development in "on-the-fly" computing technology, it becomes very essential to secure the system from unauthorized users and hackers. The biometrics verification system is one of the technological advancements in security measurements to restrict the unwanted users from accessing the systems. A behavioral biometric recognition and verification system performs the functions of reading unique biometric identity of an individual, such as fingerprint, iris, voice, and face, and authenticates the individual by comparing with the biometrics of the same kind in the database. As the above said behavioral biometric value can never be the same for two persons, the biometric verification and authentication system is highly accurate and reliable. The voice recognition and authentication system is one of the familiar and upcoming system that is taken for discussion in the chapter. To add more accuracy and intelligence flavor to the biometric especially voice-based verification and authentication, deep leaning is adopted as a pre-verification filter to separate out and remove bad or malicious individual from accessing the system. Deep learning puts core architectural building blocks together to build a new voice-based verification specific architecture. Deep learning allows the system to train itself again and again with test samples and makes the behavioral voice-based verification and authentication system (BVVAS) to be more secure and achieves high level of accuracy by reducing false identification. The deep learning supported BVVAS appears to be obviously fit for handling the ever-rising biometric identification problems ranging from mobile device to high-end security systems authentication.

Keywords: BVVAS, authentication, deep learning, finger print, iris, voice recognition

R. Regan, Computer Science and Engineering, University College of Engineering Villupuram, India, e-mail: reganr85@gmail.com
P. Rajaram, Information Technology, Malla Reddy Institute of Engineering and Technology, Hyderabad, e-mail: dr.rajaram@mriet.ac.in
K. Senthilkumar, Information Technology, Malla Reddy Institute of Engineering and Technology, Hyderabad, e-mail: kksent@gmail.com
N. Khadirkumar, Computer Science and Engineering, Maha Bharathi Engineering College, India, e-mail: khadir_n@yahoo.co.in

https://doi.org/10.1515/9783110750584-004

Introduction

Communication plays a vital role in our actual life routine, and people try to communicate with one another in a variety of ways [1]. However, human voice is more significant than all other ways for communication. All other kinds of interaction-based communications are built on the foundation of talking and listening. You would not need an electronic smart gadget to talk or listen, and you do not even need notebook and pen. You only really need your voice. Neither form of communication seems to be as easy or as straightforward [2].

Voice is a kind of human biometrics that is used for talking and listening in a communication. The voice may be used for biometric authentication since each person's voice includes unique features and specific to that voice alone [3].

Over the last decade, the same idea has substantially improved the state of the art in the field of artificial intelligence (AI). Ease of access and greater exposure to high powered, fast computing, and increased investment from major corporations have done wonders toward unlocking AI's potential, but neuroscientists are still responsible for one of our most significant breakthroughs. Artificial neural networks (ANNs) is a class of machine learning (ML) techniques inspired by the multilayered architecture of the brain of the human [4, 5].

ML is a powerful technique in which a computer simulates human brain and decisions are taken based on collected data. When a user observes an item, for example, the brain learns the features of the item and later recognizes the object based on the traits it has learnt. Deep learning is a new branch of ML research that aims to bring the AI field closer to one of its primary goals: Deep learning is a groundbreaking new ML and AI method [6].

Deep learning is the process of learning many layers of representation and abstraction to aid in the understanding of data such as pictures, sound, and text. In deep learning, the machine learns the characteristics of an item and performs a recognition job. In a variety of applications, it has set new performance standards. Image processing and voice recognition are the study fields that gets the most benefit from deep learning technology [7, 8].

Deep learning is also influenced by neuroscience and has numerous applications in handwriting recognition, computer vision, natural language processing (NLP), and voice recognition. Deep learning has greater connections to our understanding of the intelligence of the brain than traditional learning methods (learning from unlabeled input). One of the most sophisticated and promising areas of ML is deep learning, notably in the field of neural networks [9].

Deep learning is well-suited to efficient image identification processing, but it has also been employed in voice recognition applications like Amazon's Alexa or voice-to-text technologies. Deep learning is widely utilized for voice recognition because of its versatility and predicting capability, as well as its ease of use. When the world's greatest Go player was defeated by DeepMind's AlphaGo in 2016, an achievement initially believed well beyond the capabilities of even the most advanced computer systems, neural networks, gained global attention. Deep learning is believed to be applied to everything from facial recognition to automatic voice recognition (ASR) [10].

Neural networks, pattern recognition, feature learning, unsupervised learning, graphical models, optimization, and signal processing have all been interacted with and are intimately connected to deep learning. Furthermore, by employing an appropriate training method, it is feasible to increase the system performance. Deep learning is used in applications like that as search engines, Facebook, and iPhone to search for images activities [11, 12].

Deep neural networks (DNNs) use a greater number of hidden layers to learn complicated tasks, giving the network its "depth." Fewer layers may be smart enough to learn complicated functions with the same number of factors as "deep" models in some circumstances. As a result, deeper networks are not required for all purposes. Multiple layers provide the benefit of being able to learn features at different levels of abstraction. DNNs, in general, need a large amount of data to train. If the volume of data utilized to train the network is insufficient, the network may not be able to generate accurate results under test settings [13, 14].

DNNs are progressively finding use in blind feature extraction and speaker recognition/identification in state-of-the-art ASR systems. DNNs are multilayer perceptron (MLP) networks that have been trained utilizing massively parallel hardware like graphics processing units. DNNs have been effectively used to identify real-world human voice data.

As millions of consumers use those applications every day, collecting enough data for training is not an issue in these instances. However, for a voice recognition application comprising a closed set of a small number of users and a restricted training database, DNNs with a sophisticated architecture may not be required. It is feasible to utilize a DNN with less parameters in such an application.

Voice recognition, image recognition, speaker recognition, and other biometric systems employ mathematical calculations, pattern matching, and deep learning, and deep learning technology is replacing conventional approaches like hidden Markov models (HMM) used for voice recognition. DNNs have lately been utilized with remarkable success in a variety of fields. Deep learning has been used to design open-source toolkits such as Kaldi (Microsoft's flexible voice recognition toolkit), Tensor Flow (Google's software uses to build and train neural networks), and CNTK (a Microsoft-developed computational network toolkit).

Neural Network Architectures for Voice Recognition

The fundamental class of DNN topologies are deep belief networks (DBNs). This generative probabilistic model has multiple hidden layers above the output, with one visible layer at the bottom. Through the connections to the lower layers, each hidden layer unit learns the statistical representation. The complexity of the representations increases with layer height. The learning process was the main issue with DNN topologies because the standard gradient descent approach frequently fails miserably and renders DBN training virtually impossible.

A greedy layer-based unsupervised pretraining can indeed be utilized to get around the difficulty. Following the pretraining, effective supervised learning may be achieved using a technique known as fine-tuning, which employs the well-known gradient descent. The beginning weight levels for the real supervised training phase are influenced by the pretraining phase. In reality, it performs far better than traditional random weight initialization, resulting in the avoidance of local minima when employing gradient descent in back propagation. DBNs are made up of several layers of limited Boltzmann machines stacked on top of one other.

A deep Boltzmann machine (DBM) is a Markov random sector (undirected probabilistic graphical designs) with numerous levels of concealed unknown parameters. It is a symmetrically paired network of stochastic binary units. There are a number of reasons why one should use DBM designs. DBM gains from the capacity to acquire sophisticated and abstract internal representations of the input in tasks like object or voice recognition, using a little amount of labeled sensory data to perfect representations built on a vast amount of unlabeled sensory input data.

A restricted Boltzmann machine (RBM) is made up of two layers, one accessible and the other concealed. The concealed level of the lowest RBM becomes apparent to the top when two RBMs are stacked on top of each other. The objective is to use many layers of RBMs to model (represent) the input distribution as closely as feasible to reality. Multiple level layers of nonlinearity are used to achieve this aim, resulting in the extraction of a more accurate probabilistic representation for the input. The number of layers in the schema as well as the quantity on each tier are just samples. So far, it is seen that the DBN's pretraining phase is accomplished via a greedy bottom-up pass.

Convolutional neural networks are feed forward networks that use three architectural ideas to achieve shift and distortion invariance: a convolution layers field, weight sharing, and spatial or temporal subsampling. Because of the sharable weights, all feature maps at the same level are created using the same modeling; spatial and temporal subsampling (usually spatial subsuming) could possibly reduce the extracted features at the previous level by aspects, resulting in a new feature set map on which the convolution process could be performed.

Neural Network Architecture

There are some important processes in neural network architecture as follows:
- A collection of processing units is the core characteristic of a neural network.
- A network of interconnections
- A method of computation
- A process for training

Units of Processing

A neural network is made up of a large amount of really basic processing units, similar to neurons in the brain. All of these units are active at the same time, allowing for massive parallelism.

These units are responsible for all computations in the system; no other processor supervises or regulates their work. Each unit merely evaluates a scalar function from its local inputs at any given time and broadcasts the result (known as the activation value) to its neighbors. Input units receive environmental data (such as raw sensory information); hidden units, which may inwardly modify the data representation; and/or output units, which indicate decision choices or control signals.

Connections

The units in a network are arranged into a particular topology by a set of connections, or weights, shown as lines in a diagram. Each weight has a real value, which typically falls between – and + but can occasionally be constrained. While a negative weight makes one unit hinder another, a positive weight causes one unit to stimulate another. Weights can be two-way but are often one-way (from input units to output units) (mostly when there is no difference between input and output units).

The weights serve as a representation of the network's long-term memory or knowledge since their values predict how each computational algorithm will respond to each possible input sequence. Weights can alter as a result of training, although they typically do so gradually due to the accumulation of information over time.

On the other hand, activation patterns are momentary results of recent input, making them a type of short-term memory. There are three hidden layers used, each with 100 units, 91 input units, and 43 output units.

Computing

Sending an input sequence to the network or clamping an activation signal on the input units is always the first step in computing. Then, depending on the scenario, the activations of all remaining units are calculated either synchronously (all at once using a parallel system) or asynchronously (one at a time, in either arbitrary or regular order). This process, which propagates from the input layer to the output layer, is known as forward propagation in layered networks and spreading activation in unstructured networks.

Training

Training a network, in its broadest sense, is changing its connections so that it displays the proper computational behavior for all input patterns. The most typical approach is to change the weights (move the hyperplanes/hyperspheres), but it is also possible to alter the network's topology by adding or removing connections (adding or deleting hyperplanes/hyperspheres). Given that a network with multiple connections can be trained to set any of those weights to zero and have the same effect as deleting them, weight modification appears to be more general than topological alteration in certain aspects. On the other hand, topological adjustments can increase the applicability and speed of learning by restricting the model's learning parameters. Adjusting the rate of learning and momentum can help to control the parameters.

Procedure for Training

Perceptrons are the most basic form of feed forward network that uses supervised learning for training. The layers of a perceptron are made up of binary threshold units. MLPs can potentially learn each and every function, but training them is more difficult. As the hidden layer has no targets, the delta rule could not be employed directly to MLPs.

If an MLP uses linear rather than nonlinear activation functions (i.e., sigmoids instead of threshold functions), partial derivatives as well as the chain rule can be used to calculate the impact of any weight on the output activation, indicating how to alter that weight to start reducing the network's error. Back propagation is the term for the extension of the delta rule. Back propagation, which stands for "backward propagation of flaws," is a popular technique used to train ANNs that are sometimes combined with an optimization approach like gradient descent.

Automatic Speech Recognition System

Voice recognition is creeping into our daily lives. It is in our phones (Siri), game consoles (Kinect), wearables (Apple Watch), and even our houses (Apple Watch and Amazon Echo). But, given that voice recognition has been around for decades, why is it only now becoming popular? The reason for this is that deep learning has finally improved voice recognition to the point where it can be used outside of controlled situations.

The capacity of a machine or software to recognize words and phrases in a language and transform them to a device readable representation is known as speech recognition. Various voice recognition technologies currently exist, including voice calling, basic data entry, and voice-to-text conversion. Many diverse disciplines, such as pattern classification, communication theory, digital signal processing, combinatorial maths, and linguistics, are used to create ASR systems. Speech recognition is taking the place of more conventional methods of computer communication, such text input through a keyboard.

Why ASR?

Such a method is extremely beneficial to the blind in terms of providing password credentials to access any protected application. Speech recognition software can be used by those with physical impairments who are unable to use a keyboard or mouse. People with chronic illnesses that need them to restrict or limited utilization of the keyboard or mouse, like repetitive stress injuries, are also in need of voice recognition software. Individuals with cognitive and learning impairments who need to utilize voice rather than type will benefit from the speech recognition technology.

Evolution of Voice Recognition Systems

Voice recognition, at its most basic level, turns sound waves into individual letters and, eventually, phrases. The variety in sound made for the same word given accents, or cadence, is a major challenge in accurately transcribing the correct syllables (e.g., hi versus haaii).

The capacity of a machine or software to recognize words and phrases in a language and transform them to a device readable representation is known as voice recognition. Traditional means of communicating with a computer, such as textual input through a keyboard, are being replaced by voice recognition.

A good system may either replace or decrease the reliance on traditional keyboard input. In the 1950s, attempts were made to develop ASR systems. These early voice recognition systems attempted to recognize voice using a set of grammatical and syntactical rules. The system could detect spoken words if they followed a specific set of rules. Human language, on the other hand, contains a lot of exceptions to its own norms. Accents, dialects, and mannerisms may dramatically affect the way words and sentences are uttered. As a result, a deep learning algorithm is employed to accomplish ASR.

The initial instance of voice recognition occurred in 1952, when Bell Laboratories developed the "Audrey" system, which identified numbers pronounced by a single voice. In the early 1960s, IBM created "Shoebox," a machine that could identify and react to 16 words in English, along with numerals from 0 to 9, and was considered the "founder of today's modern voice recognition systems."

Advanced technologies and powerful processors, in the subsequent years, contributed to the growth and widespread usage of voice recognition applications such as Dragon Dictate. Voice recognition technology has been progressively improved over time, resulting in Apple's decision to introduce Siri in 2011.

Existing Methods

HMM and Gaussian mixture model (GMM) were previously utilized in voice identification systems. To forecast HMM states, the applications that employ an ANN use a single layer of nonlinear hidden units. Complex statistical models are used in existing ASR systems. HMMs have shown to be highly effective. Statistical models are used to produce a series of symbols or values.

Natural Language Processing

While NLP is not really a correct method for voice recognition, it is a branch of AI that deals with human communication through language, such as text to voice. Many smart mobile phones have voice recognition built in to do voice searches (e.g., Siri) or to improve messaging accessibility.

Hidden Markov Models

HMMs are based on the Markov chain model, which says that the probability of a particular state is determined by its present state rather than its past states. HMMs allow us to incorporate hidden events, such as part-of-voice tags, into a probabilistic model. They are used in voice recognition as sequence models to give labels to each item in the sequence, such as words, syllables, and phrases. With the supplied information, these labels construct a mapping, allowing it to find the most relevant label sequence.

Because a voice signal may be regarded as a piecewise static signal or a brief static signal, GMM-HMMs are employed in voice recognition. GMM-HMMs are very popular since they can be given training automatically and are simple to use and computationally viable. GMM-HMMs, on the other hand, make a variety of hypotheses about the voice and so fail to generalize.

The following are some of the drawbacks of GMM-HMM:

- It is costly, both in terms of memory and computation time.
- For modeling data that are on or near a nonlinear manifold in the data space, GMMs are statistically inefficient.
- The HMM must be trained on a collection of seed sequences, and a bigger seed is usually required.
- There are numerous potential HMMs for a given set of seed sequences, making selecting one challenging.

N-Grams

This is the most basic form of language model (LM), in which sentences or phrases are allocated probability. An N-gram is indeed a collection of N words. "play the ball" for example, is indeed a trigram or 3-gram, whereas "play the ball on merit" is a 5-gram. To enhance identification and precision, syntax and the chance of specific word sequences are employed.

Artificial Neural Networks

Neural networks, which are mostly used for deep learning algorithms, analyze data for training by simulating the interconnectedness of the brain of human using layers of nodes. Weights, input, output, and a bias make up every node. Supervised learning makes neural networks learn the mapping function, and then change it depending on the loss function using gradient descent. Despite neural networks being more accurate and handling more input, it exhibits quite a lower performance efficiency than classic LMs since it consumes longer time to train.

Mel-Frequency Cepstral Coefficient

Pitch, pitch jitter, and spectral characteristics like mel-frequency cepstral coefficients (MFCCs) and perceptual linear predictions coefficients are all present in the voice signal. These are some of the capabilities available in DNN for network training. The most crucial thing to keep in mind about voice is that it is filtered via the vocal tract, which is made up of the tongue and teeth. The form affects the sound that is produced. If the shape is properly established, it becomes possible to accurately represent the phoneme being generated. MFCC is used to break the signal up into little chunks and make the periodic signal using a hamming window. The power spectrum's periodogram estimation is also carried out using MFCC.

Speaker diarization (SD): SD algorithms recognize and partition voice based on the identification of the speaker. The process aids programs to distinguish the persons in a discussion and is often used in call centers around the country.

Voice Recognition Technique

Voice recognition technique uses voice-processing analysis to validate a person's identification. Voice authentication is a kind of voice recognition and identification. It is just not the same as speech recognition, which is utilized in speech-to-text software. Speech recognition can understand spoken words, but it cannot confirm the speaker's identification based on his or her distinctive vocal characteristics;

For starters, it is a really natural human communication interface. It is apparent that it is a fantastic method to communicate with machines because it does not require a mouse or keyboard. Most people learn to talk as a result of natural growth; therefore, you do not even need to acquire new skills. It is a really natural way to communicate with basic technologies like automobiles, chatbots, and cell phones.

Voice recognition applies to a device's capacity to accept and understand spoken commands; to put it in another way, voice recognition systems can communicate with and react appropriately to human commands. Voice recognition began on PCs, but as cellphones grew more integrated in our lives and become popular, it has become even more reachable in the usage of home devices. As a result, technology is now available and helpful in the living areas, kitchens, and other locations of our home. The popularity of voice recognition is raising as more people using it these days – this is one of the reasons why companies are considering using it. Voice Recognition Voice recognition system types are represented in Figure 1.

In today's world, there are two forms of voice recognition:

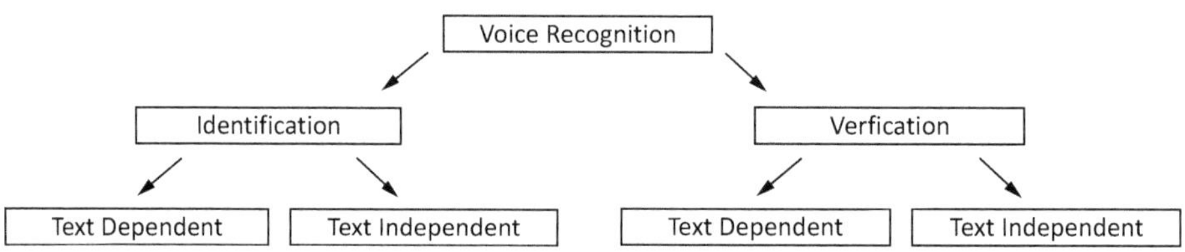

Figure 1: Type of Voice Recognition System.

- Text dependent: It works based on the precise collection of words said by the person. The system helps to do authentication and proof of identity when the user must speak the required phrase in order to make the system enabled. For both enrolment and verification, the same passwords are utilized. A speaker will be required to speak a preset phrase rather than whatever they want to authenticate. The same key is used for each verification in static text voice authentication. A randomized key, such as an alphanumeric code, is created for the participant in dynamic text-dependent voice authentication. That alphanumeric code must be registered as well.
- Text independent: The system works depend on conversational dialogue rather than a specific text. The user is not needed to speak a set of predetermined words in order to be authenticated. Any uttered passphrase or other speech material is used for voice authentication.

To generate a reference template to check with sampling for subsequent verification efforts, a speech biometric sample must be collected and registered using a microphone. After that, distinct vocal characteristics such as duration, intensity, dynamics, and pitch are examined.

The terms "speech recognition" and "voice recognition" are not identical. Only words are recognized by speech recognition, which focuses on verbal-to-text translation. Voice recognition guarantees that an individual user's voice is recognized and authenticated. By studying the rhythmic pattern and sequence of frequencies of a voice of a specific person, voice recognition is possible to verify the person and therefore authenticate the individual person. This method assures security since all inputs will be authorized exclusively by your own voice and not by anybody else's. Voice patterns are much influenced by the form and position and motion of the jaw and mouth, as well as smooth holes and airways.

It is essential to look back to see where the technology come from in order to comprehend where the technology today. Talking and listening machines have always stimulated the interest of computer scientists, partly due to the belief that an extemporaneously competent computer is a sign of true intelligence, and partly due to their practical utility. However, until recently, the efforts necessitated a Frankenstein strategy that was delicate, heavy, and difficult to scale.

Development has been difficult due to the complexities involved in human voice. It is one of its most difficult fields of computer science, including languages, arithmetic, and statistics. The voice input, extraction of features, feature vectors, a decoding process, and a word output are all elements of a voice recognizer. In order to identify the proper output, the decoder uses acoustic models, a dictionary for pronunciation, and LMs.

Voice scientists used to build distinct models to manage specific portions of the voice recognition process prior to deep learning for voice recognition and related applications. An acoustic model converted analogue sound waves into digital information that a computer could understand, a LM extracted words from that data, and a pronunciation model dealt with differences between speakers with various accents and dialects.

In the last several years, deep learning research has outperformed expectations in terms of academic influence and industry impact. Deep learning is utilized in a variety of disciplines to achieve many levels of abstraction, such as sound, image feature extraction, and text, to name a few.

Deep learning for voice recognition helps to cope these obstacles by allowing to train a single end-to-end (E2E) model that encompasses the whole processing pipeline. The attractiveness of E2E ASR designs is that they can just accept an audio input value and produce a textual result, in which all the elements of the system are trained simultaneously toward a certain objective of handling pipeline lot easier.

Voice recognition with deep learning solves many of the problems that were experienced in previous ASR iterations while also providing other benefits. Voice recognition with deep learning helps in quickly finishing the training and would not require as many resources to execute, which opens up additional deployment options. The system can also recognize dialects, accents, and different languages considerably better. Supervised learning, which trains the neural network using labeled input, is one of the finest methods for voice recognition. To teach a model to distinguish between a hen and a parrot, for example, it is essential to show photos that a human has categorized as either hen or parrot. For ASR, it entails putting the model through its paces with audio and the appropriate ground truth transcript.

Finally, while deep learning is still in its early stages in terms of voice recognition, it is rapidly approaching state-of-the-art performance. In terms of latency and accuracy, model engineering still has a lot of space for improvement. Many research breakthroughs have not been easy to put into production, therefore improvements in voice recognition must take industrial maturity into account.

Aspects of ASR

Typing is Slower than Talking

More products and devices are using voice recognition techniques to build life simpler, as voice inputs are substantially faster than typing. According to the University of Stanford, voice recognition technology has progressed to the point where it can now be considerably faster and more accurate with text outputs (e.g., voice enabled commands on a smart phone) than a person can be while typing on a keyboard. The organizations can smartly simplify administrative procedures if certain technology is used.

Personalization

Voice recognition technology is the ideal method to provide more personalization to your products and services by allowing your consumers to effortlessly and quickly convey their demands. In today's market, clients who are technologically sound and voice recognition technology would provide a special personal touch to their requirements. These intimate interactions may be generated with voice intelligence, which might improve the company–individual relationship. Even though people cannot see the speaker, everybody is trained to identifying whose voice they are hearing. If the people know the individual, they have a good chance of guessing right.

Performance and Productivity

When it comes to the business, voice recognition, such as Amazon's Alexa, can give huge support and care by extending the assistance to strategic operations and management responsibilities such as establishing teleconferences, organizing appointments and meetings, and putting up reminder notifications. The organizations may use this technique to increase the efficiency and effectiveness, resulting in increased production and performance. With the advancement of voice recognition, it has become feasible to gather useful information in response to a voice command. All of these management tasks may now be completed in less time than they would take if done manually.

Simple to Use

Because voice technology simply records a person's voice, it is an excellent way for people with motor impairments or disabilities to communicate with your business more easily. People who previously used technology slowly or not at all now have access to it thanks to technology. Furthermore, using voice recognition software can help people with a variety of disabilities, such as arthritis or hand tremors, which can be worse by too much typing during the day.

Multiple Users' Access

Unlike before, when a customer service representative could only chat to one individual at a time, voice recognition system allows them to contact several consumers at once. Businesses may expand the number of clients they serve and deal with various inquiries or issues more effectively using the technology. Because voice technology is always evolving and getting more advanced, businesses are increasingly seeking new methods to utilize it in order to expand their reach and increase their business outcomes.

Hands-Free

Voice recognition may help to complete tasks more quickly no matter what you are doing or what profession you are doing. Talking to a customer, taking notes after a client meeting, or communicating to a coworker about a meeting held can take quite a long time when done manually. Voice intelligence recognition system may speed up the activities by eliminating the need for extra documents as all customer information, notes, and queries will be captured automatically and properly.

Voice recognition technology seems to be a trend that many individuals and organizations seek to implement since it may help both homes and enterprises. Business owners started to accept utilizing voice searches and voice assistants in their media and communication strategy as the technology became trendier. The progression and advancement of voice technology is still in progress, but keep in mind that having voice assistance in your daily life has gradually confirmed to be beneficial.

Architecture of ASR

The following are the processing steps in a typical ASR system:
1. Preprocessing
2. Extraction of features
3. Classification
4. LM and authentication

By lowering the signal-to-noise ratio, decreasing noise, and filtration of the signal, the preprocessing stage tries to enhance the quality of audio signal. In practice, the ASR features are obtained with a set of values or coefficients that are created by employing various techniques to the input. The stage has to be reliable in terms of quality issues like noise and the echo impact.

The following extraction of features approaches are used by the most of ASR techniques:
- MFCCs
- Wavelet transform (discrete)

The most popular approach for obtaining voice characteristics is MFCCs. The human ear is a nonlinear system that receives audio signals. The mel scale was created to provide a linear model of the human auditory system in order to cope with the shift

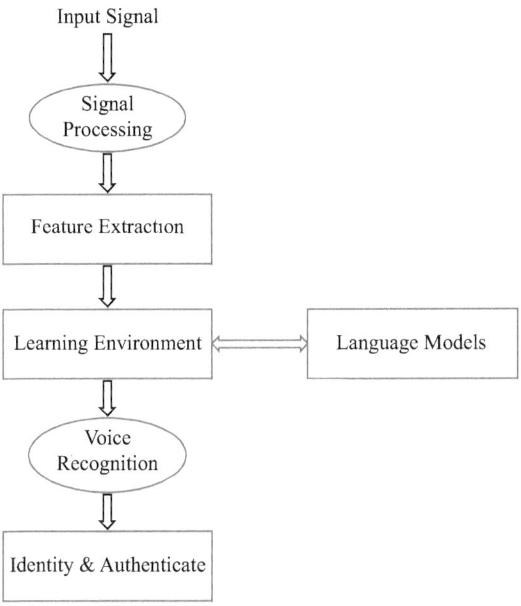

Figure 2: Architecture diagram of automatic voice recognizing system.

in frequency. Only frequencies between 0 and1 kHz ranges may be converted to mel scale, whereas all other frequencies are considered logarithmic.

The classification model's goal is to locate a spoken word on the input signal. It creates the output using the retrieved features from the preprocessing stage.

LM is considered as a crucial element since it incorporates a language's grammar rule structure and semantic relation information. LMs are so crucial for interpreting the output token from the classification model as well as making text corrections if any.

From Figure 2 Physical elements, including the forms of the vocal chords and the oral cavity and speaking patterns are specific to each one of us and determine the distinctive qualities of every person's voice. By separating and evaluating the elements that correspond to these particular physical and behavioral attributes, voice recognition technology may identify the speaker.

Speaker identification or speaker authentication are the technical terms for speech recognition. In several situations, it relates to the technique that compares two sounds using one-to-one matching analysis to see if they are the identical person. Speaker identification, on the other hand, uses one-to-many analysis to determine the unknown person based on their speech. Even yet, it all comes down to a series of one-to-one comparisons repeated numerous times.

The most often used technology for feature extraction is i-vector. i-vector captures the variations between standard version and the supplied voice as a feature using a standard model of phonemes composed of numerous speakers' voices. If all

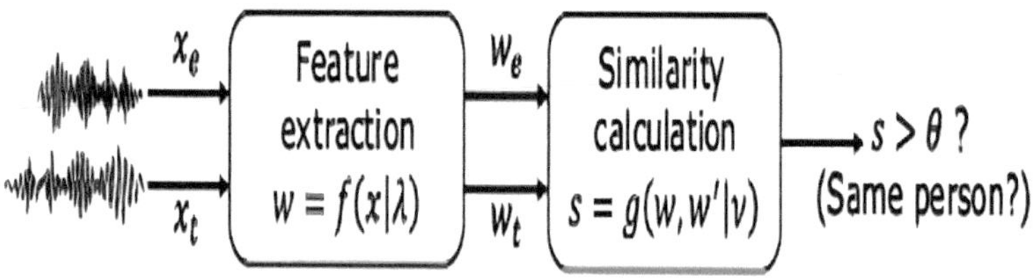

Figure 3: Basic configuration of voice recognition system.

of the variations are removed, the feature will be huge, with large numbers of dimensions. To overcome the issue, i-vector uses component analysis to reduce such a large feature to about 400 dimensions. A technique known as probabilistic linear discriminant analysis (PLDA 2) is frequently used to assess similarities. Based on the 400-dimension feature of the i-vector, the PLDA randomly constructs equations using linear discriminant analysis (LDA), a classic approach for ML, and automatically picks the feature particularly fits for speaker identification. From Figure 3, The similarity is estimated as a probability ratio once the data has been examined.

Both i-vector and PLDA are constructed on the prediction of Gaussian function and are developed using prediction model. Various ML approaches are incorporated into the illegitimate protocols for speech recognition, i-vector, and PLDA. They can create appropriate parameters of the model from a vast quantity of data via automatic learning.

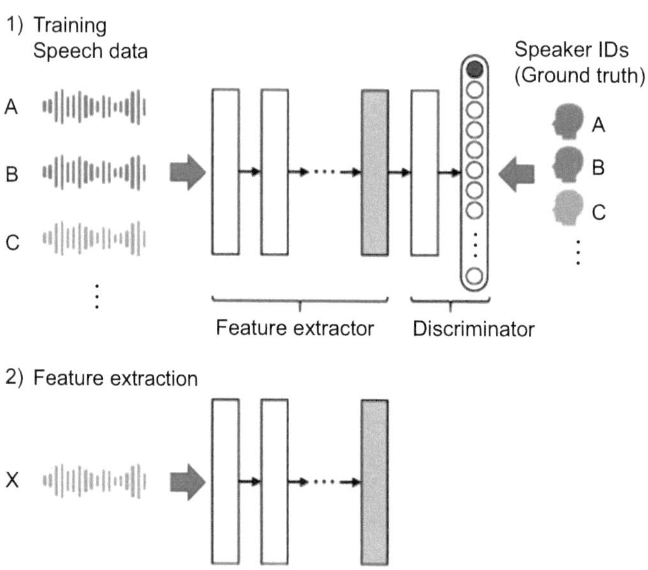

Figure 4: Feature extraction and training.

The introduction of a technology known as deep speaker embedding, or x-vector, which enhances the accuracy of speech recognition tenfold, marks this transition. Research studies in the area have taken the opportunity to use x-vector as a major feature extraction technique that might potentially replace the i-vector system.

The quantity of data supplied to the neural network is likewise changeable since voice is data of time series with different lengths. Because of this, dealing with voices seems to be more challenging than dealing with visuals. From Figure 4, By introducing a convolution layers – which accumulates the input in a temporal direction – at the conclusion of the feature extractor, it is feasible to produce a feature in a predetermined set of measurements using the x-vector.

The use of deep learning in voice recognition goes beyond feature extraction and includes everything from the front end (voice recognition and voice improvement in noisy environments) to the back end used for similarity comparison.

ASR Datasets

Text from different sources like audio books, chats, and talks has been collected and stored in several databases. The conversation data in the CallHome's English, Spanish, and German databases include a substantial number of nonvocabulary words. Foreign language and telephone channel distortion provide a challenge to databases, making it difficult to keep them up to date.

There are 120 random English phone conversations in the English CallHome database between fluent English speakers. The training dataset has 80 dialogues totaling about 15 h of voice, while the test and development sets each comprise 20 dialogues totaling 1.8 h of audio clips.

Furthermore, the CallHome Spanish includes 120 phone chats with fluent Spanish speakers. The training section contains 16 h of voice, while the exam section has 20 dialogues totaling 2 h of voice. Eventually, the CallHome German is made up of 100 phone calls between native German speakers, with 15 h of voice in the training set and 3.7 h in the test set.

TIMIT2 is a huge dataset of American English broadband recordings in which each participant speaker reads 10 semantically rich phrases. TIMIT includes time-aligned, rectified audio signals that can be utilized for conventional character or word identification. The audio clips are 16-bit encoded. The training set comprises audios of 462 different speakers, while the validation dataset contains audios from 50 different speakers and the test dataset contains audios from 24 different speakers.

Characteristics of a Successful Voice Recognition System

Although there are many voice recognition devices and systems available, the most cutting-edge alternatives rely on AI and ML. The systems may mix grammar, syntax, structure, and audio and voice signal composition in order to understand and analyze human voice. The systems should, in theory, pick up new information so they may respond differently to each contact.

The best solutions also help organizations modify technology to suit their particular requirements, from brand recognition to subtle language and speech changes. Consider the following scenario:

- Language weighting: In addition to terms currently in the standard vocabulary, improve accuracy by weighting key phrases that are uttered frequently (such as product names or industry slang).
- Speaker labeling: Create a transcription of a multiparticipant discussion that references or identifies every speaker's input.
- Acoustics training: Pay attention to the business's acoustics. Train the device to adjust to different speaking styles and acoustic environments (such as those seen in call centers) (like voice pitch, volume, and pace).
- Profanity filtering: Utilize filters to detect and clean the unwanted things to make better voice output by identifying specific words or phrases.

However, voice recognition technology is improving. The big corporates are making progress in a number of aspects to improve human–machine interaction.

The Applications of Voice Recognition System

Hands-free smartphone authentication is the most common use for voice authentication. This is great for mobile phones and other situations where facial recognition, fingerprint recognition, and other kinds of biometric identification are cumbersome, such as in cars.

Speech-recognition systems like Google Home and Amazon's Alexa can also benefit from voice authentication. Virtual assistants are increasingly being utilized to conduct orders and carry out other tasks that need identification. During customer-service conversations, speaker recognition can also be used as an authenticator. Users may find this easier and secure than giving personal details for identification validation, such as their driver's license or credit card information. Voice technology is already being used in a broad range of sectors, allowing businesses and consumers to save time and even lives. Here are a few examples:

Automobiles

Voice recognizers in vehicle radios increase driving safety by allowing voice-enabled navigation systems and search tools.

Healthcare

The doctors use voice recognition software to collect and record patient diagnoses and treatment notes in the healthcare industry.

Sales

Voice recognition technology may be used in a variety of sales scenarios. It can assist a call center in transcribing numbers of client and dealer phone calls in order to explore typical call patterns and concerns. AI chatbots may also communicate with consumers through a website, addressing frequent questions and completing simple requests without the need to wait for a contact center employee. In both cases, voice recognition technology helps minimize the time it takes to resolve customer concerns.

Mobile Phones

Virtual agents have become more and more incorporated into human's daily lives, particularly on mobile devices. Voice commands are used to control mobile devices via applications, such as Google Assistant or Apple's Siri, for activities like voice search. Voice recognition applications are also used in mobiles to playback the music through our speakers using Amazon's Alexa or Microsoft's Cortana.

Security

As technology becomes more integrated into our daily lives, security procedures are becoming more important. Signatures are no longer necessary for modest credit card purchases. By simplifying and accelerating the payment process, obstacles in purchasing for both consumers and sellers are lowered. Consumers nowadays value ease just as much as security. Voice recognition satisfies both of these requirements. Because speech is a common medium for daily communication, biometric authentication via voice offers users with a convenient and straightforward method of identifying themselves.

Individual authentication in business activities such as e-commerce and online banking is made possible via voice recognition.

Advantages and Disadvantages

Several significant advantages of voice recognition over other kinds of identity authentication include:

Since all cellphones possess microphones, it is readily available for authentication on mobile phones. Integration with other devices, like as vehicles and home appliances, is more economical. For the most part, consumers will find it appropriate and comfortable. Because it is contactless, it is less intrusive and sanitary.

The following are some of the downsides of voice recognition:

Other biometric techniques are more accurate (e.g., facial recognition). To ensure that a sample comes from a live speaker rather than a recording, liveness detection is required. Noise in the background might affect sample quality and, as a result, degrade the matching performance. It is not suitable for many situations like noisy environment or public areas.

Conclusion

Voice recognition is unquestionably one of the most straightforward biometrics to set up and utilize. This technology is now considerably more dependable and safe owing to the inclusion of deep learning in speech recognition systems. With excellent technology that sets the bar for accuracy and performance, voice recognition is suitable for a variety of applications, including e-commerce, contact centers, and police case, since it is consumer friendly and accurate. Several research studies are providing technical advantages to humanity, as well as efforts are underway to improve and perfect it.

References

[1] G. Salazar, A. Romero, and D. Juge, Personal communication, Human Computer Interface, Avionics Systems Division, NASA Johnson Space Center, Houston, TX, 2016.

[2] D. Povey, A. Ghoshal, G. Boulianne, N. Goel, M. Hannemann, Y. Qian, P. Schwarz, and G. Stemmer, "The Kaldi Speech Recognition Toolkit." IEEE Signal Processing Society, Hilton Waikoloa Village, Big Island, Hawaii, US, 2011.

[3] V. Ramakrishnan, P. Chenniappan, R. K. Dhanaraj, C.-H. Hsu, Y. Xiao, and F. Al-Turjman, "Bootstrap aggregative mean shift clustering for big data anti-pattern detection analytics in

5G/6G communication networks," in Computers & Electrical Engineering Elsevier, (vol. 95, p. 107380), 2021.

[4] D. Garcia-Romero, X. Zhang, and A. McCree, "Improving speaker recognition performance in the domain adaptation using deep neural networks," in IEEE Spoken Language Technology Workshop, South Lake Tahoe, NV, USA, 2014.

[5] F. Richardson, D. Reynolds, and N. Dehak, "Deep neural network approaches to speaker and language recognition," IEEE Signal Processing Letters, vol. 22, no. 10, pp. 1671–1675, 2015.

[6] G. Jiuxiang, Z. Wang, J. Kuen, and L. Ma, "Recent advances in convolutional neural networks," arXiv.org, Pattern Recognition, Volume 77, Pages 354–377, 2018, 2015.

[7] L. Krishnasamy, T. Ramasamy, R. Dhanaraj, and P. Chinnasamy, "A geodesic deployment and radial shaped clustering (RSC) algorithm with statistical aggregation in sensor networks," Turkish Journal of Electrical Engineering & Computer Sciences, vol. *29*, no. 3, pp 1464–1478, 2021.

[8] S. Han, H. Mao, and W. J. Dally, "Deep compression: Compressing deep neural networks with pruning, trained quantization and huffman coding," ICLR, https://doi.org/10.1007/s00521-022-07161-0 2016 Vol.no: 9, pp no: 67–73.

[9] G. Nijhawan and D. M. K. Soni, "Speech recognition using MFCC and vector quantisation," International Journal on Recent Trends in Engineering and Technology, vol. 11, no. 1, pp 211–218 2014.

[10] M. Denll, B. Shaklbl, L. Dlnh, M. A. Ranzato, and N. D. Freitas, "Predicting parameters in deep learning," Advances in Neural Information Processing Systems, vol 26. pp 1–9, ARXIV, 2013.

[11] D. R. Kumar, T. A. Krishna, and A. Wahi, "Health monitoring framework for in time recognition of pulmonary embolism using internet of things," Journal of Computational and Theoretical Nanoscience, vol. 15, no. 5, pp. 1598–1602, 2018.

[12] L. Prechelt, "Early stopping – But when?," in Neural Networks: Tricksof the Trade, Springer, Berlin, Heidelberg, pp 53–67, 2012.

[13] Y. Dong and L. Deng, Speech Recognition – A Deep Learning Approach, Microsoft Research, ISBN 978-1-4471-5779-3.

[14] A. B. Downey, Think DSP Digital Signal Processing in Python; Joe Tebelskis, Speech Recognition using Neural Networks, May 1995, CMU-CS-95-142.

Oleeviya Babu P, Sobin C. C, Jahfar Ali

Building Machine Learning–Based Prediction System for Critical Diseases

Abstract: When new technologies are created for the welfare of the humans, it also brings many challenges, particularly when it applied to healthcare. Machine learning is one of such new technology which is implemented to solve many of the problems in healthcare. Machine learning techniques have a huge impact on today's day-to-day activities. Most of the applications are going to be automated using such techniques. Also, we are living in an era of providing better healthcare services to the human using technologies. In this chapter, we performed a study on role of machine learning techniques in healthcare applications and built prediction system for some of the critical diseases like sickle cell anemia, breast cancer, and heart diseases. We have analyzed some of the biomedical data with existing machine learning algorithms and identified some possible directions to the research community.

Keywords: Healthcare, machine learning, sickle cell disease

Introduction

The wide usage of computer technologies in the medical fields increase the amount of electronic data. The supervised algorithms help the medical experts to understand and identify the symptoms in the early stage. In this chapter, we choose three diseases such as sickle cell disease, heart disease and breast cancer for the prediction of the chances of these diseases using different machine learning algorithms. k-nearest neighbor (KNN), random forest (RF), decision tree, gradient boost classifier, support vector machine (SVM), naive Bayes model, and logistic regression (LR) are the machine learning techniques that are used for the prediction of these critical diseases. The emergence of artificial intelligence (AI) makes the computer behave like a human. This will help our world achieve many successes. One of them is in the field of medical industry. The aim of this chapter is to predict the occurrences of the critical disease in the early stages and thereby help improve the status of the medical field. The following sections help to know more about the three critical diseases.

Oleeviya Babu P, Otto von Guericke University, Magdeburg, Germany,
e-mail: oleeviyazehr@gmail.com
Sobin C. C, Department of CSE, SRM University AP, India, e-mail: sobin.c@srmap.edu.in
Jahfar Ali, Language Technologies Research Centre, IIIT Hyderabad, India,
e-mail: jahfar2004@gmail.com

https://doi.org/10.1515/9783110750584-005

Sickle Cell Anemia

Sickle cell anemia (SCA) causes the red blood cells in the patient's body to change their shape and texture to a crescent or sickle cell shape and rough, which happens due to the abnormalities caused due to the abnormalities in oxygen-carrying proteins in red blood cells. This causes trouble in blood flow compared to the normal blood cells. The detection of SCA is prominent in younger ages since the symptoms are visible at an early age. They may sometimes appear in infants at 4 months old, but generally occur at 6 months old. The manual assessment and classification require a lot of time. With the help of different data-mining techniques like classification algorithms, sickle cells in the human body can be effectively classified with short time and high accuracy and it also helps to identify in which condition the person is. Whether the person is a carrier, sickle cell patient, and if there is no crisis (controlled condition).

There are so many drawbacks of manual assessment, and the proposed method overcomes it with the help of classification algorithms thereby classifying the SCA in blood cells into two classes, such as sickle cell patient and sickle cell trait. In this chapter, we analyzed the accuracy of various classification algorithms by performing training in the dataset. The dataset was created based on the clinical and laboratory information provided by the primary health center (PHC) located at Wayanad, Kerala, India. We compared the blood cells of a normal person and a sickle cell patient. From Figure 1, we can see that the shape of the normal RBC is a biconcave disc whereas the shape of the sickled RBC is like a crescent moon. In the case of blood flow, let us analyze another figure.

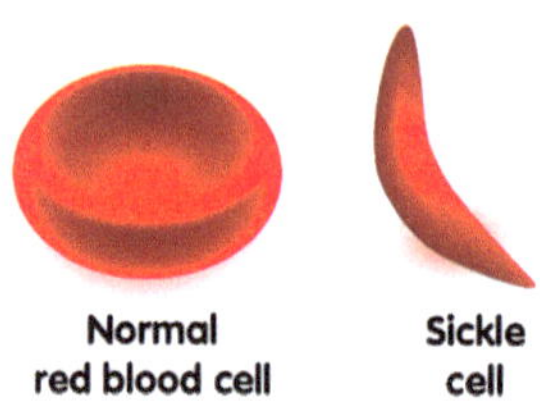

Figure 1: Comparison of normal RBC and sickled RBC [1].

From Figure 2, it is clearly evident that the normal blood cells carry oxygen to all parts without any blockage of the blood, whereas sickle cell shape will stop the blood flow at any time because it is sticky. Sickle cell disease is inherited from a couple to their children if both of them have the condition, which is sickle cell disease. In some cases, the children may have sickle cells without having any troubles or with traits. If a person is having sickle cell disease if both of his/her parents have the disease condition. The person is called a carrier if he/she received sickle cell gene from only one parent.

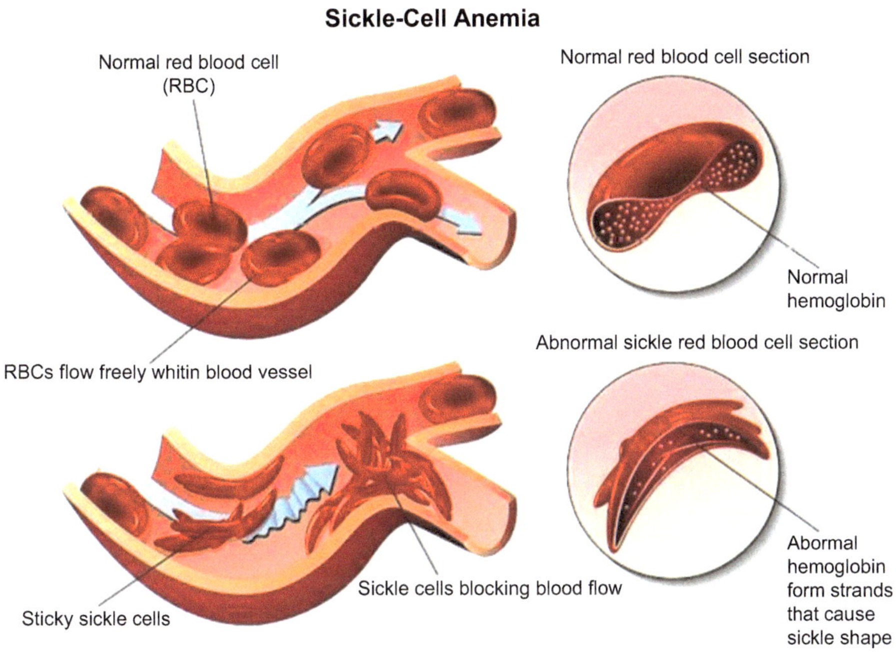

Figure 2: Comparison of normal RBC and sickled RBC [2].

SCA can lead to some complications, some of them are:

A. Acute chest syndrome
The condition in which sickle cells blocks the blood vessels in your lungs. It leads to many critical issues including chest pain and possible death [3].

B. Pulmonary hypertension
It is a disease that affects adults; it causes high blood pressure in the lungs. Fatigue and shortness of breathing are the common symptoms [4].

C. Stroke
When blood flow to the brain is blocked, it causes stroke. Symptoms of stroke include loss of consciousness, sudden speech difficulties, numbness in arms and legs, seizures, and so on [5].

D. Organ damage
SCA also causes blockage of oxygen-rich blood to different organs which in turn leads to organ damage [6].

Different methods can be used for the detection of the SCA. Slide method, electrophoresis, and analysis of blood parameters are the methods used for the detection of sickle cells in our local PHC. Slide methods (Figure 3a) and electrophoresis (Figure 3b) mainly are used. EDTA blood is added to the slide and observed. If there are 25% or more HbS present in the RBCs, then the test is positive. The next level is to do electrophoresis in which current is passed through the blood and hemoglobin which have different charges migrate to different positions. Based on the position or the percentage of the types of hemoglobin present in the blood, we can determine whether a person is sickle cell patient, sickle cell trait, and normal patient. Let us look at the figures of both methods.

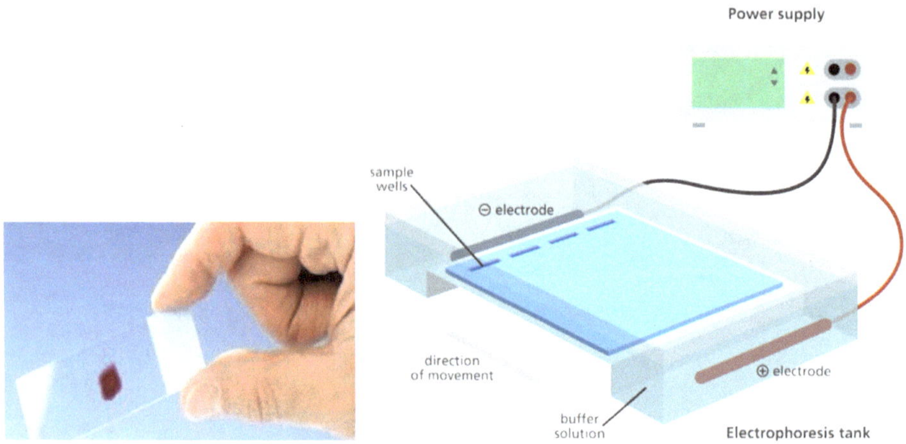

Figure 3: (a) Slide method [7]. (b) Electrophoresis method [8].

Heart Disease

The condition which affects our cardiovascular system is called heart disease. One of the common diseases is heart attacks. The common reason for this disease is the blockage of the blood vessels and the coronary arteries that pump blood to the heart itself. The objective of this chapter is to analyze the different types of symptoms of the heart disease by previously used data and thereby using the dataset to predict the occurrences of the disease in new patients. Different symptoms are heart valve disease, diseased heart valve muscle, heart infection, and so on [9].

An average human heart beats approximately 100,000 beats every day and pumps around 2,000 gallons of blood. The signs of heart disease in women are less noticeable than in male. Range of conditions that affect the heart is called heart disease. It contains different types such as [9]

- Heart valve disease
- Disease of the heart muscle
- Herat infection
- Heart rhythm problems

Electric wiring of the heart makes it beats and the heartbeat controls the exchange of the blood and thus we live. The cause for heart disease is based on the type of disease. Heart has four valves and it helps the flow of the blood by closing and opening. There are many reasons for the damage of heart valves [9] . Some of them include:
- Breathing problems
- Fatigue
- Chest pain
- Irregular heartbeat

Apart from the above listed reasons, heart disease symptoms due to diseased heart muscle are not visible at the early stages [9]. However, when the condition worsens, it includes symptoms such as:
- Swelling of the legs
- Fainting
- Dizziness
- Irregular heart beats

Heart infection is an infection that affects the heart valves and heart chambers, and the symptoms can include:
- Weakness
- Swelling in abdomen
- Swelling in legs
- Fever
- Dry cough
- Irregular heart beats

There are some existing research works in the area of predicting heart disease using machine learning techniques. Based on the previously obtained data, we can make predictive models for the prediction of cardiovascular diseases. In this chapter, we have analyzed 14 attributes from a set of 74 attributes and performed the classification using the machine learning techniques, such as LR, decision tree, and RF, and obtained the results.

Breast Cancer

Breast cancer is a type of cancer that develops in the breast cells. It can occur in women and rarely in men. The cancer is typically developed in the ducts or lobules of the breast. Fatty tissue or the fibrous connective tissue of the breast are also affected by the cancer cells. These cancer cells destroy the other healthy breast cells and also travel through the lymph nodes, Lymph are the main pathways that help to move the cancer to the other parts of the body [9].

Early symptoms are very difficult to identify because the tumor is very small to feel but can be identified using a mammogram. There are different varieties of breast cancer but most of the symptoms are almost similar, but few were different. Some of the common symptoms are given as follows:
- Breast lump and tissue thickening
- Breast pain
- Peeling of breast skin
- Bloody discharge from the nipples
- Discharges from the breast other than the milk
- Unexpected change in the shape of the breast

Types of Breast Cancer

Different kinds of breast cancer are classified under the main two groups such as invasive and noninvasive [10].
- *Invasive lobular carcinoma*: As the name suggests, it comes under the invasive group. It is first developed at the lobules of the breast and spread to the nearby areas or the tissues.
- *Invasive ductal carcinoma*: This type also comes under the invasive category. It is a commonly seen breast cancer. It first develops in the breast and then spreads to the other parts of the tissues of the breast.
- *Lobular carcinoma in situ*: It comes under the category of the invasive. It first develops in the milk-producing glands and does not get invaded to the nearby glands.
- *Ductal carcinoma in situ*: It comes under noninvasive category, and it is developed only in the ducts of the breast and does not invade nearby tissues.

Analysis of Breast Cancer Stages

By analyzing the parameters and stages of breast cancer, we can predict whether a person has breast cancer or not. Breast cancer can be categorized into stages based on the size of the tumor. Breast cancer has five stages. In this study, we will analyze each and every stage, and then predict whether the person has cancer or not. In order to understand the stage of breast cancer, the doctor needs to know some important points, whether the cancer is invasive or not, whether the lymph nodes get involved, size of the tumor, and if the cancer will spread to the nearby tissues.

STAGE 0:
- It is the stage in which the cancer cells are only in the duct and not get spread to the nearby tissues of the breast.

STAGE 1:
- Stage 1A: The stage in which the tumor is 2 cm wide and will not affect the lymph.
- Stage 1B: The stage in which the tumor is 2 cm wide and also the lymph is affected.

STAGE 2:
- Stage 2 A: In this stage, the tumor is less than 2 cm, and it spreads to 1–3 lymph nodes. Otherwise, the tumor is about 2–5 cm large, and the lymph is not affected.
- Stage 2 B: The tumor is between 2 and 5 cm large and also affects the 1–3 armpit node or the tumor is larger than 5 cm and the lymph is not affected.

STAGE 3:
- Stage 3 A: In this stage, it will affect 4–9 axillary lymph nodes and also the mammary gland gets enlarged. The tumor can be of any size.
- Stage 3 B: It is the stage in which the tumor invaded the chest walls and the lymph may or may not get affected.
- Stage 3 C: Ten or more axillary lymph nodes are infected by the cancer cells.

STAGE 4:
- This is the stage in which the cancer can be of any size and also it affects many lymph nodes and also the distant organs in the body.

In this chapter, we analyze different attributes of breast cancer from the previous data and perform the classification using the machine learning algorithms. Breast cancer has different types of symptoms and some of them are breast pain, peeling of the skin, bloody discharge, and so on. The datasets of the breast cancer consist six attributes, and we analyze each attribute and perform the classification for the prediction of disease in the new patients using the machine learning techniques. The machine learning algorithms used for the prediction are LR, SVM, KNN, decision tree, and gradient boost classifier.

Literature Review

In this section, we will describe the existing literature for SCA, heart diseases, and breast cancer.

Sickle Cell Disease

This section contains the analysis and the examination of the existing methods related to this field that contains different information using different techniques. We analyzed the existing literature based on the following factors.
- The intensity or severity of the sickle cell disease
- The machine learning algorithms used
- The correct dosage of the hydroxyurea

Most of the research papers that are included in this literature part contain the prediction of the intensity of the disease and also about the various kinds of risk factors. The remaining includes the usage of hydroxyurea along with the implementation of the machine learning techniques.

Ademola et al. [11] discussed the usage of ensemble models in the prediction of the severity of the sickle cell disease among the children. They have used seven different types of machine learning models for the prediction of the disease. It also includes the features of the laboratory results. The classifiers used for the prediction are C4.5 DTs, naive Bayes, SVM, and ensembles of the isolated classifiers. The models were created using the 10-fold cross validation. The best result is obtained from the ensemble method that was formed by the combination of the C4.5 DTs and naive Bayes classifiers. The authors provided an information about the relation of the demographic and clinical features with the severity of the sickle cell disease.

Abd Al-Maaref et al. [12] discussed about a health monitor intelligent system to monitor the patient's health. The system is installed in the patient's smartphone and the patient can check about their health condition on their own. Whenever the patient updates all the information regarding their health like the blood tests, the method will decide about the critical aspects of the disease. It will check and give suggestions to the person to take care of the routine and daily medicines for a person who is not in a critical solution. On the other hand, for a person who is not in good condition, or in a critical condition, the system will directly suggest approaching a doctor. The main purpose of the system is to provide 24/7 service to the needy. The dataset used to train the classifiers include patients who are traits and patients who are not traits. This paper performs various kinds of machine learning techniques including the Jrip rules, attribute selected classifier, voted perceptron, Bayes Net, Adaboost.M1, and LogitBoost. The authors claim that the LogitBoost gives the best accuracy compared to other algorithms.

Khalaf et al. [13] developed an intelligent system for sickle cell disease using smartphones. The objective is to overcome the challenge faced by the patients. A challenge that many patients face is to maintain proper communication with their doctor. The intelligent system helps to overcome this problem. When the patient provides information about their condition, the intelligent system tells whether the person is sickle cell trait or not a carrier. This classification uses several machine learning algorithms which provide good accuracy in the prediction. The only difference between the first mentioned paper and this is the usage of different classifiers. The multilayer perceptron gives us the best accuracy in the prediction of whether a person is critical or not.

Khalaf et al. [14] have discussed the importance of pain management among sickle cell patients. The authors have mentioned that, if the assessment is accurate, all the steps in the pain management can be done with correct accuracy. But it is true that the pain is a subjective matter and cannot express it using objective methods. The authors provide a management system by drawing objective physiological measures from the subjective experience using the machine learning techniques. They have analyzed the data of 40 individuals, and multinomial LR is used to predict the patient's pain with best accuracy. They have also employed other machine learning techniques such as SVM, KNN, and RF. The pain is measured in a 11-point scale. It is from 0 to 10, 0 means no pain and 10 means unbearable pain. The pain prediction is based on different situations. From all the above situations, accuracy, and F1 score, the authors claimed that the multinomial LR provides the best accuracy when compared with the remaining algorithms.

Mohammed et al. [15] discussed about the prediction of the organ failure among sickle cell patients who were hospitalized using the machine learning technologies. The authors have collected data from 63 patients having sickle cell with 163 encounters. The authors have employed various machine learning classifiers including the MLP, SVM, RF, and LR. By using these technologies, early detection of organ failures is possible and can implement various care methods and protocols to the person suffering. This will also help the patients to prepare their mental health. By identifying organ failures, implementation of the early healthcare is also possible. The results obtained from the experiments shows that RF predicts the organ failure with greatest accuracy.

Xu et al. [16] developed a deep neural network architecture that used convolutional neural networks to classify the heterogeneous shapes in the affected blood and to use independent shape analysis which results in the use of dual approaches to detect sickle cell disease. Bhavnani et al. [17] used a photo imaging microscope is to extract geometric and texture features of the red blood cells using image processing techniques. Optimization technique and mean filter plays the crucial role. The SVM is used to classify red blood cells data as either normal or abnormal, that is, it finds out if the patient has sickle cell or not. Solanki et al. [18] used classification techniques like J48 and random for the prediction of sickle cell disease and comparison of the two techniques have taken place.

Heart Disease

In this section, existing literature on machine learning techniques used for prediction of heart diseases are listed.

Santhana et al. [19] proposed heart disease prediction using machine learning algorithms. The authors mainly focused on the prediction of heart disease among males, and it explained the risk factors, facts, types, and problems of the heart disease. Nagaraj et al. [20] predicted the heart diseases using naive Bayes classification and SVM. Based on the analysis, SVM is the superior algorithm when compared to the naive Bayes classification.

Mohan et al. [21] also used machine learning techniques for the prediction of heart diseases. The main objective is to predict the critical and also improve the exactness of the expectation of the cardiovascular technology. The accuracy level of the improved expectation model is 88.7%. The improved expectation model is created using the hybrid RF with a linear model.

Kishore et al. [22] used deep learning techniques for the prediction of heart attack possibility. The authors have also used recurrent neural system are used for the prediction of the possible heart infections and provided the details about the framework related to the hypothesis.

Nagamani et al. [23] proposed a method for heart disease prediction using data mining techniques along with MapReduce algorithm. The model in this paper is more accurate than the conventional fuzzy artificial neural network. Dynamic schema and linear scaling are the two techniques used in this data mining which makes the model more accurate.

Breast Cancer

Breast cancer is developed in the breast cells and is diagnosed when there is an abnormal lymph in the breast, or a small speck of calcium is found on the X-ray. When it is found, the doctor will again diagnose whether it is cancerous or not. In this section, existing literature for breast cancer prediction is summarized.

Chauhan et al. [24] proposed genetic algorithm-based ensemble approach for the breast cancer prediction. They have used techniques such as decision tree, RF, SVM, neural networks, linear model, AdaBoost, and naive for the prediction. The limitation of classical weighted average method is overcome by the usage of the new technique called the genetic algorithm based weighted average ensemble method. Finally, based on the experimental results, authors concluded that genetic algorithm performs better.

Azar et al. [25] proposed a method for breast cancer detection using variants of decision tree, such as single decision tree (SDT), boosted decision tree (BDT), and decision tree forest. During the training phase, SDT and BDT have an accuracy of

97.07% and 98.83%, respectively, and from these results, it is concluded that BDT is better than SDT. Keles et al. [26] used data mining classification techniques for prediction of breast cancer. The main objective is to detect the tumor using the data mining algorithms at the very early stage. Waikato Environment Knowledge Analysis tool is used for the comparison of different data mining algorithms. The dataset used in this paper consists of 6,006 values and among those values 5,405 is used as the training dataset and the remaining 601 is used for testing. In order to get the most authentic results, 10-fold cross-validation technique is used. After the testing using the 10-fold cross-validation, RF performs best and provides an accuracy of 92.2%.

Li et al. [27] proposed using machine learning techniques for prediction and also evaluated the performance of the models. Two types of datasets were used in their study. After performing the training and testing using the two different datasets on five different models, their accuracy is compared based on the F-measure and ROC curve. Based on the performance, it is concluded that RF gives the best accuracy. But there are also some limitations in their study. The proposed method consists only limited attributes and this can be solved in the future by using more research and survey on the patients. The limited data may have some impact in the final result.

Proposed Method

In this section, we will describe about the three critical diseases such as SCA, heart diseases, and breast cancer, and the prediction models.

Sickle Cell Disease

In case of sickle cell disease, the challenge we faced was the lack of data. We went to the PHC in Wayanad district to collect data, but there are only 30 data (12 years old data) which is not sufficient. Based on the discussion with the doctors, health inspectors, and health nurses, based on the available data, features, and information given by the health community, we created an artificial dataset for our initial study purpose. We studied and analyzed the results of slide method and electrophoresis and based on those details we create the attributes of our dataset. The attributes are hepatitis B surface antigen (HbS), fetal hemoglobin (HbF), hemoglobin A (HbA), hemoglobin A2 (HbA2). Based on the range of each attribute, we can predict weather a person is sickle cell disease patient or sickle cell trait. Slide method alone cannot predict whether a person is sickle cell or sickle cell trait. Slide method only says weather a person has sickle cells. So, we need to analyze the results of the electrophoresis.

After analyzing the electrophoresis result and comparison of the range of each parameter with the normal range, we create the dataset. Further, we are studying about the possibilities of the reduction of time and efforts required for the detection of SCA. The data was synthesized in a uniform manner by following the guidelines from the clinical experts. In total, 1,510 data points were generated in which 950 negative samples and 560 positive samples were there. Train–test split on the data was done on the proportion of 60–40, that is, 60% training data and 40% testing data.

Data preprocessing is the first stage. In the next stage, the given data is split into train and test sets respectively. After training the machine learning model, the performance of the model is evaluated. After the splitting, the training phase and testing phase took place.

In this section, we analyzed the classification algorithms. These given machine learning algorithms are the models that used to make a separation between the different kinds of objects based on certain rules. Further, we discuss models used in our study.

Logistic Regression

LR, at its very form, is a linear method which consists of logistic function. The logistic function can be used to vary the predictions we obtain from the models. The impact of this methodology lies in the fact that we can no longer interpret the predictions with linear combinations as we do in linear regression analysis. It can be clearly said that this algorithm is a linear method, and it consists of a logistic function so the predictions can be varied according to the Logistic function [28].

Random Forest

RF is an ensemble of many random decision tree classifiers which is being used for solving both the classification problems and regression problems. The results are finalized according to the voting of individual classification trees. The RF model is efficient, in both the sense of handling large datasets and in providing accurate results. It contains a plethora of classification trees. They are one of the best, effective, and accurate general-purpose techniques which are used for learning as well as classification. It can perfectly handle the bigger datasets, so it will be easier to perform the training and testing if we have larger datasets. It has the capability to understand the importance of variables and can manage the outliers and make it fit for the training [29].

KNN Model

KNN implements a case-based learning method which is non-parametric in nature and an effective method for classification tasks. KNN, even though it is non-parametric, is not a complex model; in fact, it is simple to implement. It maintains all the training data for classification [30].

SVM

SVM is a supervised learning method which is associated with learning algorithms that analyze the datasets for classification as well as regression. SVMs are extensively being used in machine learning tasks because of its effectiveness in classification of data points. The SVM algorithm tries to find out a hyperplane which is best suited in order to classify the data points [31].

Decision Tree

Decision tree has a tree like structure in which the branch of the tree depicts the output of the test we input, and each leaf of the tree indicates the class labels of the input. The decision tree algorithm makes use of the decision tree structure to generate probabilistic results regarding an item [32].

Naive Bayes Classification

Naive Bayes classification is based on Bayes principle, and the classifier is built on top of a cluster of classifiers in a way such that each classifier is independent. The naive term is being used because the classifier at first makes unsophisticated assumptions from which the model performs. It is a group of algorithms where each and every one shares a basic and common principle [33].

Gradient Boosting Classifier

In gradient boost algorithm, on each iteration, a new model is added to the previous ones in a way such that the model learning is done by minimizing the error that occurred on previous model runs. This method can be used both for classification problems and regression problems. This algorithm consists of a loss function and the loss function is getting minimized [34].

Heart Disease

In order to identify whether a person has heart disease or not, we need to collect the data of those who have heart disease previously and also at the same time we need those who do not have heart disease. The datasets used for the prediction of the heart disease are taken from the website called Kaggle. Python language is used for programming the experiment. The dataset used was the UCI dataset and it consists of a total of 74 attributes. Many experiments were taken before using these datasets, for all the early held experiments a subset of 14 attributes was used. Therefore, we are also choosing the already processed dataset. The major task is to find whether a person has heart disease or not based on the given attributes. The heart disease datasets are listed in Kaggle [35].

Breast Cancer

In order to predict whether a person has breast cancer or not, we need to collect the data of those who have breast cancer previously and also at the same time we need those who do not have breast cancer. The datasets used for the prediction of the breast disease is taken from the website called Kaggle, we found a public dataset from the Kaggle. The dataset consists of six columns [36]. All the features have floating point values except the diagnosis. We perform the machine learning techniques in the breast cancer dataset and get the following accuracy results. We perform these algorithms on the datasets and predict whether a person is a cancer patient or not. Based on that prediction, we create an accuracy table to show which algorithm performs the best.

Experimental Results

This section clearly analyzes and calculates the performance of machine learning classification algorithms on the dataset of three critical diseases.

Sickle Cell Disease

To reach the success and the ultimate objective of this research, some training or machine learning techniques have been done on the sickle cell disease dataset that we have created. These classification algorithms are implemented by using Python Jupiter notebook with Tensor flow.

Different machine learning algorithms are evaluated and compared for the diagnosis of sickle cell disease. The given dataset is split into training data (60%) and testing data (40%). By analyzing the table of accuracy (Table 1), F1_score (Table 2), precision (Table 3), recall (Table 4), and ROC curve (Table 5), it is clearly understood that LR, SVM, and Gaussian Naïve Bayes are the best performers. KNN is the second-best performer for sickle cell dataset. Decision tree is the least performer.

Table 1: Sickle cell dataset accuracy comparison.

Method	Accuracy
Logistic regression	76.18
KNN	75.84
SVM	76.18
Naïve Bayes classification	76.18
Decision tree	58.93
Random forest	74.65
Gradient boosting classifier	72.86

Table 2: Sickle cell dataset F1_score comparison.

Method	F1_score
Logistic regression	86.66
KNN	86.46
SVM	86.66
Naïve Bayes classification	86.66
Decision tree	74.05
Random forest	86.12
Gradient boosting classifier	85.78

Table 3: Sickle cell dataset precision comparison.

Method	Precision
Logistic regression	89.21
KNN	88.78
SVM	89.21
Naïve Bayes classification	89.21
Decision tree	75.48
Random forest	89.10
Gradient boosting classifier	89.44

Table 4: Sickle cell dataset recall comparison.

Method	Recall
Logistic regression	84.25
KNN	84.25
SVM	84.25
Naïve Bayes classification	84.25
Decision tree	72.68
Random forest	83.33
Gradient boosting classifier	82.40

Table 5: Sickle cell dataset ROC comparison.

Method	ROC
Logistic regression	89.29
KNN	89.16
SVM	89.29
Naïve Bayes classification	89.29
Decision tree	79.77
Random forest	88.83
Gradient boosting classifier	88.49

By applying these common machine learning algorithms, we had discussed previously, to the synthetic dataset on sickle cell anemia, and came to a set of conclusions. Among the different machine learning techniques applied, naive Bayes algorithm performed well compared to others. The decision tree classifier showcased the worst performance on a test set. The dataset is synthetic, and the results are dependent on how we synthesized our data. As described in earlier section, the data was built on top of data we received from medical professionals. The procedure the medical professionals are currently using is almost error-free looking from a data perspective. So, in general, there is a huge possibility that the models we define on top of such precise data will be well-performing under any given scenario. But, for even more realism, we did consider the factor of human error while preparing the data. The factor we considered is in the worst-case scenario, but in the real world, there is no scope or very little scope for errors, both in data and in the model as there is a clear relation between variables in this scenario. The graphs given in Figure 4 help us to clearly understand the best performer classifiers.

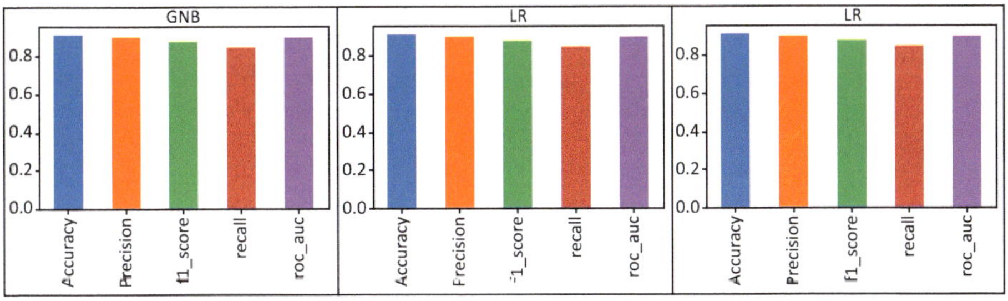

Figure 4: Comparison of algorithms.

Heart Disease

The dataset mentioned in Section "Heart disease" under Section "Literature review" is split into 80% for training and 20% for testing. We have used algorithms such as RF, LR, and decision tree for the prediction of heart disease. The efficiency of the algorithms is calculated based on the different metrics and the results are given in Table 6.

Table 6: Comparison of heart disease results.

Algorithm	Precision	F-score	Recall	Accuracy
Decision tree	0.842	0.825	0.835	81.87%
Random forest	0.936	0.901	0.883	90.1%
Logistic regression	0.856	0.868	0.881	85.20%

From Table 6, we can clearly conclude that RF gives best result comparing to the LR and decision tree. An accuracy of 90.1% is produced by this model. The second-highest accuracy is given by LR. With the increasing number of deaths due to cardiovascular disease, it is necessary to create efficient and accurate models for the prediction of the heart attacks, heart infections, and other heart related diseases. Based on the analysis, RF gives the best among the three. Figure 5 represents the efficiency of the techniques.

Breast Cancer

The dataset mentioned in Section "Breast cancer" under Section "Literature review" is split into 80% for training and 20% for testing. The attributes in the dataset are passed as the input to the algorithms such as KNN, SVM, gradient boosting classifier, LR, and decision tree. The results of the breast cancer prediction are given in the Table 7.

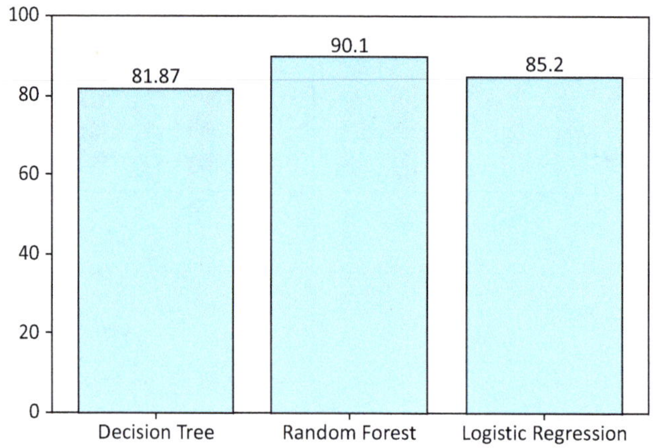

Figure 5: Accuracy of algorithms.

Table 7: Comparison of breast cancer results.

Techniques used	Accuracy
KNN	87.6
Logistic regression	89.06
SVM	90.45
Decision tree	91.78
Gradient boosting classifier	94.0

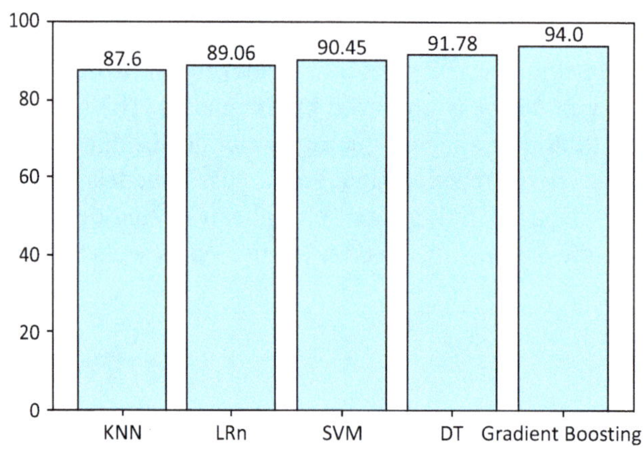

Figure 6: Accuracy of algorithms.

From Table 7, it is clearly evident that the gradient boost provides the best accuracy for the breast cancer prediction, which is about 94%. Decision tree stands in the second position, which has an accuracy of about 91.78%. SVM provides an accuracy of 90.45%. LR stands in the fourth place (89.06%) and KNN has the least accuracy which is about 87.6%. From this accuracy table, we can understand that the gradient boosting classifier gives better results compared to others. The graphical representation of the accuracy of the breast cancer is given in Figure 6.

Conclusion

Sickle cell disease is a common disease now. It will be a big problem to the people who suffered. It requires immediate attention and care for the patients and also awareness. This chapter presents a description about the disease and methods for detection which takes a lot of time to discover the disease. Naive Bayes classification, LR, and SVM gives better results for classification. KNN algorithm gives the second-best accuracy for the prediction.

One of the challenges we faced during the implementation stage was the lack of real-life data regarding the problem. The data records are very little in quantity and are hard to find. The input data records are statistically under clear boundaries, and if we manage to overcome the factor of human error while collecting the data, the method discussed will be error free. In the future, we are interested in expanding the search for dataset regarding the problem statement and see how well the model performs in real-life data. The proposed methodology can be used to provide an accurate diagnosis for sickle cell anemia.

Application of machine learning algorithms for heart disease prediction will make noticeable changes in the field of healthcare. New types of method can be drawn from the existing methods and can make more and more methods in the future. The machine learning techniques we have used in this chapter perform best with the available attributes of the heart disease and give us the result. The machine learning techniques learn and understand the previous data and predict the disease in the early stage which helps the medical field and also protects the life.

The breast cancer prediction using machine learning techniques is also a great challenge in the medical field. Our experimental results show that more techniques can provide more accuracy. Performing classification using different types helps us to find which will give more efficiency than others. New types of method can be drawn and can make more methods in the future. The techniques used in this chapter perform best with the available attributes and give us the result which helps us to identify the disease in the early stages.

The list of abbreviations used in this chapter is listed in Table 8.

Table 8: List of abbreviations.

AI	Artificial intelligence	SVD	Sickle cell disease
KNN	k-Nearest neighbors	SVA	Sickle cell anemia
ANN	Artificial neural network	SVM	Support vector machine
PHC	Primary health center	RBC	Red blood cells
EDTA	Ethylenediaminetetraacetic acid	HbS	Hepatitis B surface
BPNN	Back-propagation neural network	WEKA	Waikato Environment for Knowledge Analysis
CNN	Convolutional neural network	NN	Neural network
MAE	Mean absolute error	SSE	Sum of squared error
RMSE	Root mean squared error	PSO	Particle swarm optimization
BDT	Boosted decision tree	DE	Differential evolution
DTF	Decision tree forest	HbF	Fetal hemoglobin
ROC	Receiver operating characteristic	GA	Genetic algorithm

References

[1] https://urbanleagueupstate.org/prisma-health-to-combat-sickle-cell-disease-with-comprehensive-program/ (Accessed on 14 February 2021).

[2] https://www.tamingthesru.com/blog/intern-diagnostics/laboratory-evaluation-of-sickle-cell-disease-in-the-ed (Accessed on 14 February 2021).

[3] Chest syndrome, https://www.texaschildrens.org/health/acute-chest-syndrome-children-sickle-cell-disease (Accessed on 14 February 2021).

[4] S. Yaghi, A. Novikov, and T. Trandafirescu, "Clinical update on pulmonary hypertension," Journal of Investigative Medicine, vol. 68, no. 4, pp. 821–827, 2020.

[5] Sickle Cell symptoms causes, https://www.texaschildrens.org/health/acute-chest-syndrome -children-sickle-cell-disease (Accessed on 14 February 2021).

[6] Sickle cell anemia symptoms causes, https://www.mayoclinic.org/diseases-conditions /sickle-cell-anemia/symptoms-causes/syc 20355876 (Accessed on 14 February 2021).

[7] https://study.com/academy/lesson/blood-smear-test-procedure-staining-interpretation. html (Accessed on 14 February 2021).

[8] https://www.yourgenome.org/facts/what-is-gel-electrophoresis (Accessed on 14 February 2021).

[9] www.mayoclinic.org (Accessed on 14 February 2021).

[10] https://www.breastcancer.org/symptoms/diagnosis/stagig (Accessed on 26 February 2021).

[11] B. J. Ademola, A. Temilade, E. N. Chidozie, and I. P. Adebayo, "An ensemble model of machine learning algorithms for the severity of Sickle Cell Disease (SCD) among pediatric patients," Computer Reviews Journal, vol. 2, pp. 331–346, 2018.

[12] D. Abd Al-Maaref, J. K. Alwan, M. Ibrahim, and M. B. Naeem, "The utilisation of machine learning approaches for medical data classification and personal care system management for sickle cell disease," in 2017 Annual Conference on New Trends in Information Communications Technology Applications (NTICT), Baghdad, Iraq, 2017.

[13] M. Khalaf, A. J. Hussain, D. Al-Jumeily, R. Keenan, P. Fergus, and I. O. Idowu, "Robust approach for medical data classification and deploying self-care management system for

sickle cell disease," in Proceedings of IEEE International Conference on Computer and Information Technology; Ubiquitous Computing and Communications; Dependable, Autonomic and Secure Computing; Pervasive Intelligence and Computing, Liverpool, UK, 2015, pp. 575–580.
[14] M. Khalaf, A. J. Hussain, D. Al-Jumeily, R. Keight, R. Keenan, P. Fergus, and I. O. Idowu, "Training neural networks as experimental models: Classifying biomedical datasets for sickle cell disease," in International Conference on Intelligent Computing. Springer, Cham, August 2016, pp. 784–795.
[15] A. Mohammed, P. S. Podila, R. L. Davis, K. I. Ataga, J. S. Hankins, and R. Kamaleswaran, "Machine learning predicts early-onset acute organ failure in critically ill patients with sickle cell disease," bioRxiv, pp. 614941, 2019.
[16] M. Xu, D. P. Papageorgiou, S. Z. Abidi, M. Dao, H. Zhao, and G. E. Karniadakis, "A deep convolutional neural network for classification of red blood cells in sickle cell anemia," PLoS Computational Biology, vol. 13, no. 10, pp. e1005746, 2017.
[17] L. A. Bhavnani, U. K. Jaliya, and M. J. Joshi, "Segmentation and counting of WBCs and RBCs from microscopic blood sample images," International Journal of Image, Graphics and Signal Processing, 2016, vol. 8, no. 11, pp. 32–40.
[18] A. V. Solanki et al., "Data mining techniques using weka classification for sickle cell disease," International Journal of Computer Science and Information Technology, vol. 5, no. no. 4, pp. 5857–5860, 2014.
[19] Mr. J. Santhana Krishnan and Dr. S. Geetha, "Prediction of heart disease using machine learning algorithms," in 1st International Conference on Innovations in Information and Communication Technology(ICIICT), 2019.
[20] N. M. Lutimath, C. Chethan, and B. S. Pol, "Prediction of heart disease using machine learning," International Journal of Recent Technology and Engineering, vol. 8, no. 2S10, pp. 474–477, 2019.
[21] S. Mohan, C. Thirumalai, and G. Srivastava, "Effective heart disease prediction using hybrid machine learning techniques," IEEE Access, vol. 7, pp. 81542–81554, 2019.
[22] A. Kishore, A. Kumar, K. Singh, M. Punia, and Y. Hambir, "Heart attack prediction using deep learning," International Research Journal of Engineering and Technology (IRJET), vol. 5, no. 04, pp. 2395–0072, 2018.
[23] T. Nagamani, S. Logeswari, and B. Gomathy, "Heart disease prediction using data mining with mapreduce algorithm," International Journal of Innovative Technology and Exploring Engineering (IJITEE), ISSN: 2278-3075, vol. 8, no. 3, pp. 2278–3075, January 2019.
[24] P. Chauhan and A. Swami, "Breast cancer prediction using genetic algorithm based ensemble approach", in 9th IEEE International Conference on Computing, Communication and Networking Technologies (ICCCNT), 2018, pp. 1–8.
[25] A. T. Azar and S. M. El-Metwally, "Decision tree classifiers for automated medical diagnosis," Neural Computing and Applications, vol. 23, no. 7, pp. 2387–2403, 2013.
[26] M. K. Keleş, "Breast cancer prediction and detection using data mining classification algorithms: A comparative study," Tehnicki Vjesnik, vol. 26, no. 1, pp. 149–155, 2019.
[27] Y. Li and Z. Chen, "Performance evaluation of machine learning methods for breast cancer prediction," Applied and Computational Mathematics, vol. 7, no. 4, pp. 212–216, 2018.
[28] H. Das, B. Naik, and H. Behera, "An experimental analysis of machine learning classification algorithms on biomedical data," in Proceedings of the 2nd International Conference on Communication, Devices and Computing. Springer, 2020, pp. 525–539.
[29] A. Pretorius, S. Bierman, and S. J. Steel, "A meta-analysis of research in random forests for classification," in 2016 Pattern Recognition Association of South Africa and Robotics and Mechatronics International Conference (PRASA-RobMech). IEEE, 2016, pp. 1–6.

[30] Y.-L. Cai, D. Ji, and D. Cai, "A KNN research paper classification method based on shared nearest neighbor," in NTCIR, 2010, pp. 336–340.
[31] C. Chih-Feng and M.-H. Horng, "The construction of support vector machine classifier using the firefly algorithm," Computational Intelligence and Neuroscience. 2015 (2). https://doi.org/10. 1155/2015/212719.
[32] H. Sharma and S. Kumar, "A survey on decision tree algorithms of classification in data mining," International Journal of Science and Research (IJSR), vol. 5, no. 4, pp. 2094–2097, 2016.
[33] P. Kaviani, "Mrs. sunita dhotre, "short survey on naive Bayes algorithm"," International Journal of Advance Engineering and Research Development, vol. 4, no. 11, 2017.
[34] V. Athanasiou and M. Maragoudakis, "A novel, gradient boosting framework for sentiment analysis in languages where nlp resources are not plentiful: A case study for modern greek," Algorithms, vol. 10, no. 1, pp. 34, 2017.
[35] UCI, Heart Disease Data Set. https://www.kaggle.com/ronitf/heart-disease-uci (Accessed on 26 February 2021).
[36] https://www.kaggle.com/c/breast-cancer-classification/data (Accessed on 26 February 2021).

Shahab Nadeem Hashmi, Nitesh Bharti, V. M. Manikandan

An Overview of Internet of Things and Machine Learning for Smart Healthcare

Abstract: Due to COVID-19, the attempt to access medical facilities has increased. But, physical meetings are the biggest reason for the global pandemic. The healthcare industry is in a dire state of despair because of the spread of COVID, constrained health workers, and infrastructure. Healthcare accommodations are costlier than ever due to various reasons such as the global population getting busier and the rise in the number of chronic diseases. Nowadays, there is a high demand for healthcare support due to the increase in the aging population, the COVID-19 crisis, its post-disease difficulties, and other chronic diseases. Treating all the patients at hospitals is not practical nowadays due to limited healthcare facilities. The patients' care is all dependent on visits to the hospital physically, and to some extent, calls, text messages, and emails. This way it was not possible for the doctors to monitor the health of the patients continuously and assist them whenever needed. Recently, the Internet of Things (IoT) and various machine learning approaches have been widely explored in this sector. The IoT-based healthcare methods provide quality and efficiency during treatment. The IoT along with machine learning techniques can transform the healthcare industry by making it more accessible and less costly and make healthcare more facile by equipping the users with pocket amicable medical facilities. With IoT-enabled devices, remote monitoring can enable patients to better connect with doctors, thereby minimizing physical visits. It can equip doctors to easily keep track of their patient's health by utilizing wearables like fitness bands that can monitor various healthcare parameters. All these systems mostly use machine learning or deep learning approaches. This way, they will know whether to offer any immediate medical care or to make any changes to the current treatment. This chapter gives a detailed discussion on IoT and machine learning in the healthcare sector and the security concerns in this domain along with the major research challenges.

Keywords: Healthcare, machine learning, Internet of things, remote healthcare, security, COVID

Shahab Nadeem Hashmi, Nitesh Bharti, SRM University-AP, Andhra Pradesh, India
V. M. Manikandan, SRM University-AP, Andhra Pradesh, India, e-mail: manikandan.v@srmap.edu.in

https://doi.org/10.1515/9783110750584-006

Introduction

Today's globalization is redefining how companies manufacture "things." It sets out how companies can achieve faster innovation and increase efficiencies. IoT, or the Internet of Things, refers to technologies that allow everyday consumer and industrial products to become attached to the Internet. These technologies enable them to exchange data about their actual performance, temperature, movement, and location and provide instant data analysis and smart responses. By assigning a unique identification to each object, IoT makes it possible for its users to live smarter, safer lives.

Several aspects of our society are set to be transformed by this technology. Various reports by the World Health Organization confirm that health expenditure globally has increased drastically to cover the scope of universal health coverage. The development of smart healthcare services is now receiving serious attention from at least 30 countries. Many researchers have examined how modern technologies can help detect serious illnesses like chronic psychological disorders, chronic disorders, brain tumors, and breast cancers in an early stage. It is critical that doctors, IT managers, and administrators who are responsible to provide better connectivity, a trusted network setting in modern healthcare to control smart technologies in use. Secure interoperability of healthcare infrastructure and enhanced communication between core members remain challenging. Efficient health systems are vital to handling the healthcare concerns to provide better healthcare support. The majority of the medical funds from government are allocated to healthcare to treat diseases such as tuberculosis, malaria, and reproductive health. There is an urgent need for standard healthcare facilities to enable citizens to live in safer and more comfortable conditions in a dynamic environment.

Healthcare is one area where IoT has significant impacts, as it is under the consideration to reduce the expenditure for treating unwell population. There are still a lot of paper-based processes in the healthcare devices and services industry due to a burdensome regulatory environment. By taking advantage of an increasingly interconnected and responsive world, IoT can help healthcare organizations to be more productive. The idea is to connect the ecosystem, thereby increasing interconnectedness. Healthcare practitioners can use various monitoring equipments such as sensors and smart pills to gather data continuously, and other systems can examine the data to identify the variations in a patient's health conditions and recommend treatments. This will accelerate the discovery and dissemination of new treatments.

Additionally, it allows cross-human or cross-object communication via the Internet or between objects using sensors. As a result, medical devices, sensors, diagnostic, and imaging devices constitute a core part of the IoT. A doctor will be able to monitor their patient anywhere, which will reduce the number of trips to the doctor.

History

The term IoT is a 16-year-old concept. Despite this, the idea of connected devices has been around since at least the 1970s. However, the word "Internet of Things" did not receive widespread attention for the next 10 years until Kevin Ashton coined it in 1999. The inception of the twenty-first century saw discussion of the "Internet of Things" by *Boston Globe*, *Forbes*, *The Guardian*, and other prominent media outlets. An international conference on IoT was organized in Switzerland in 2008 after the masses observed a steady increase in interest in the IoT. The conference garnered international attention as there were participants from 23 countries, and discussions on new-age devices such as short-range wireless communications, RFID, and sensor networks were done. In addition, several significant developments paved the way for IoT development such as

- Among LG Electronics products introduced in 2000 was a refrigerator with internet connectivity so users could shop online or make video calls.
- Developed in 2005, a rabbit-shaped robot "Nabaztag" provided updates on weather forecasts, prices in the stock market, and news.
- Google StreetView gathered not only 360°2019 images but also stored huge amounts of data about people's WiFi networks. There was speculation that this was the beginning of Google indexing the internet and the physical world.

The IDC published a report in October 2013 estimating that the IoT would be an $8.9 trillion market by 2020. When Google announced in January 2014 that it would buy Nest for $3.2 billion, the term "Internet of things" became popular. At the same time, the Consumer Electronics Show in Las Vegas was themed "Internet of Things." There are currently more than 27 billion connected devices, and experts expect it to rise to over 100 billion by 2030.

The intersection of interconnected technologies and various industries is also opening up new areas:

- Smart healthcare
- Smart cities and buildings
- Internet of medical things
- Automotive IoTs
- Smart retail
- Industrial IoT
- Smart agriculture

Present Scenario

IoT presents many advantages as a standalone technology, but combining it with artificial intelligence, augmented reality, and cloud computing opens up many more possibilities. Mixed solutions will become more prevalent in the future.

For example, blockchain-based IoT applications will help decentralize networks and increase the security of data transfers between interconnected devices. It is no secret that blockchain is one of the hottest trends in the IoT, and blending it with IoT will have even more benefits.

Machine learning and AI are fundamental to the future of IoT. As an example of an application, smart home devices that learn your preferences, or autonomous production processes that automate themselves, are also examples. Through machine learning, IoT devices will be able to make their own independent decisions and become more innovative by themselves rather than simply reporting data.

In 2019 and beyond, cloud and edge computing will remain critical for IoT data storage, with experts expecting that edge computing will grow in popularity.

Internet of Things and Machine Learning Approaches in Healthcare

A segment with the fastest growth rate in the IoT market is smart healthcare devices, commonly known as Medical IoT. By 2026, the value of this industry, also known as the Internet of Medical Things (IoMT), is expected to reach $176 billion. The IoT has sparked incredible breakthroughs and advancements in the healthcare industry. The IoT is changing how and where care is delivered and the quality of treatment and the patient experience. In 2020, an estimated 646 million IoT devices will be utilized by healthcare practitioners and consumers, according to a Forbes article, and that figure is predicted to expand dramatically.

With the current pandemic, Healthcare solutions based on the IoT offer the potential to improve outcomes by millions and save lives through remote patient monitoring and decentralized clinical trials – all while addressing rising healthcare costs and an aging population as well as for patients to monitor themselves.

In the past, patients had limited access to doctors outside of in-person visits, teleconferences, and text messages. Hospitals or doctors could not provide real-time information about patients' health or make suggestions.

IoT-enabled devices now make monitoring remotely in the healthcare sector, allowing physicians to provide exceptional care to patients while also protecting them from harm. The ease and efficiency of interactions with doctors have also increased patient's satisfaction. Similarly, remote monitoring of a patient's health can help reduce hospitalizations and avoid readmissions. In addition to lowering medical costs and improving patient outcomes, the IoT has a substantial influence.

Benefits of IoT and Machine Learning in Healthcare

The benefits of IoT for the healthcare industry are innumerable. Some of them are:

- **Cost-cutting:** The adoption of IoT technology can enable hospitals and healthcare professionals to minimize waste costs.
- **Error reduction:** Human error is declined as a result of IoT software automation's numerous manual operations.
- **Fast diagnosis:** Using constant patient examining through real-time data, the illness can be diagnosed in the initial stage, even before symptoms appear.
- **Proactive treatment:** Constant monitoring of health parameters allows for proactive treatment.
- **Enhanced patient experience:** Patients' experience is enhanced over time as they are more engaged in their treatment due to IoT connectivity, regardless of their location, and healthcare providers have better access to all pertinent patient data. Patient satisfaction has improved.
- **Reduced hospital admissions and readmissions:** With an always-on IoT connection, healthcare practitioners can monitor patients in real time, tracking their progress and spot problems before they require hospitalization.
- **Accurate disease diagnosis:** Since IoT devices along with machine learning approaches can collect more data points regularly, physicians can also make more precise diagnoses, sometimes even before major symptoms appear.
- **Improved drug and equipment management:** IoT makes medicine and equipment tracking easier, and providers can spend less time looking for what they need. Equipment management is a major concern in the healthcare industry. Because of connected devices, these can be controlled and exploited more efficiently and at a lower cost.
- **Assisting the elderly:** People over the age of 60 run the risk of major health problems due to unexpected falls, forget to take medications, etc. Several potential solutions to this issue are provided by the IoT and machine learning that allow the elderly to remain independent while still receiving emergency assistance. In the days following the Covid-19 pandemic, when many family members could not visit their elderly relatives, these services were essential.
- **Observing cleanliness requirements:** IoT can trigger reminders for proper sanitary measures when hospital staff enters rooms, helping to reduce infection rates. IoT can deliver specialized and customized reminders if specific cleaning processes are required based on a patient's demands.

- **Sensors that can be ingested:** Doctors can avoid invasive probes of a patient's digestive tract for some medical procedures. Ingestible devices save time and provide convenience and comfort to the patient.
- **Surgical robotics:** Surgeons can execute difficult treatments with robotic surgeries, thanks to the IoT. Patients benefit from smaller incisions, less pain, and a shorter recovery period with these aided operations. They also allow hospitals to have shorter stays with less recovery time.

Various IoT Devices in Healthcare Domain

Due to new demands on healthcare, and the need for remote diagnosis, a lot of IoT devices are in use. Here are some examples of how intelligent gadgets connected to the IoT are supporting healthcare, facilitating medical treatment, and preventing illness:

- **Hospitals and virtual wards:** By leveraging innovative technology and connected medical wearables, healthcare experts can deliver outpatient and long-term care remotely to patients at home, creating critical rooms for patients requiring more acute care. Countries such as Australia, the United Kingdom, and the Middle East have tested and used virtual hospitals or wards.

An English NHS trust, Northampton General NHS Trust, is collaborating with Doccla (a company provides virtual hospital care) to provide remote caring for Covid-19 patients and patients with other severe illnesses in July 2021.

In February 2021, Sydney opened its first virtual hospital, the RPA Virtual Hospital, just as the Covid-19 pandemic hit Australia. In a matter of hours, the hospital "remodelled" to provide remote treatment to patients suffering from Coronavirus symptoms. Various parameters such as oxygen saturation, body temperature, and heart rate are captured and sent to the virtual hospital's employees via an app on the patients' phones. Due to COVID-19, virtual hospitals revolutionized in various ways to deliver remote care.

A few devices that are in use for healthcare sector are detailed below:

- **Smart thermometers:** It uses intelligent technology while Coronavirus crisis. It was Kinsa's decision to publish anonymized, collective details taken by smart thermometer throughout the United States in a map depicting groups of high temperatures that may imply a COVID outbreak. This highlighted a new application of intelligent technology. It can indicate far broader trends and patterns relating to health and wellness on a regional, national (or even global) scale and provide data and insights on an individual level. While Covid-19 is not the only illness that can cause a high temperature, Kinsa distinguished Covid-19 clusters from typical flu and common cold epidemics by comparing data gathered during the commencement of Covid-19 to data gathered in past.

In addition to tracking infections on a broad scale, intelligent thermometers and their accompanying apps can make recommendations for treating infections by considering symptoms and age. This information is beneficial for parents of small kids, who are not understood how severe an disease is or what signs to look for. Additionally, intelligent thermometer apps collect logs that show health trends over time, so a doctor or caregiver can get a second opinion from the records.

– **Wearables biosensors:** The wearable biosensor is an essential technology that allows virtual hospitals to take care of illness avoidance and recognition. These compact, lightweight gadgets are attached on the body and observe critical symptoms such as body temperature, heartbeat rate, and breathing frequency, giving health practitioners crucial information about beginning of an illness or condition of a patient.

An experiment involving wearable biosensors found that the devices can be used both inside and outside of a clinic setting to observe patients with multiple sclerosis. In a study published in Neurology Advisor, researchers collected information from patients with cardiac activity monitoring devices, which equates to 50,000 h of data collection.

A wearable smart device shows clearly why they are so helpful in examining patients with long-term illness. Since data collection is not limited to patients' time during hospital visits, the physicians can now collect data from patients all the time.

– **Automated insulin delivery systems:** A new type of insulin delivery system called an automated insulin delivery system (AID) or artificial pancreas offers life-changing relief to people with diabetes, which affects almost 8.5% of adults worldwide. Along with these systems, continuous glucose monitors (CGMs) observe blood glucose levels to decide the quantity of required insulin. The medication is then administered using an insulin pump. CGM in an AID system constantly observe the glucose levels, even while the user is sleeping or resting. By delivering continuous insulin into their bodies through the pump, automatically adjusted based on glucose readings from the CGM, they can spend a lot more time "in range," with fewer severe blood sugar fluctuations.

In the last 7 years, DIY artificial pancreas gadgets have become more common. In 2013, a father with a Type 1 diabetes wife and child built the first automated artificial pancreas device connected to the cloud storage. Upon creating Smartloop Labs (now Bigfoot Biomedical), Mazlish expanded and commercialized his idea to create an AID system. A free and open-source device, the Open Artificial Pancreas System (OpenAPS), was invented by Dana Lewis and her husband Scott Leibrand allowed users to build their AID system. Since these gadgets are becoming more user-friendly and affordable, they might even prevent more serious health conditions in diabetics by reducing extreme blood sugar highs and lows and reducing the burden on the healthcare system, allowing people to live their lives with more stability and minimal health problems.

– **Connected inhalers:** Chronic obstructive pulmonary disease and asthma are treatable breathing conditions, but failure to follow a prescription therapy and inadequate monitoring may cause serious complications. The Royal College of Physicians found that two out of every three asthma deaths could be avoided if asthma plans were developed for patients, timely reviews of care occurred, and more appropriate drugs were prescribed.

An appealing solution to this problem in the age of smart technology has emerged in the form of smart inhalers. Like smart thermometers, connected inhalers, which are also frequently supplemented by an application, assist users with respiratory disorders in tracking their medication usage by providing reminders to take their medication on time.

– **Smartwatch monitoring:** Although smartwatches were not designed to be medical devices, a range of applications and capabilities make them a potential healthcare device. It is reported that Apple has been working to understand more about the changes in level of blood oxygen. The company has consistently committed to turning the Apple Watch into a health checking and support device. As part of the health applications built into smartwatches, users can track sleep cycle activity and use activity trackers and movement reminders to prevent unnecessary weight gain through guided breathing and meditation workouts, which are beneficial to mental health. Using handwashing applications and reminders, Apple, Google, and Samsung have all worked to encourage thorough handwashing since the outbreak of Covid-19. It could be argued that making these instruments more widely available could aid in the prevention of the onset of more severe illnesses by detecting early warning signs.

Related Works in IoT Security

There has been an exponential growth of multimedia technology and high data traffic in every industry. Similarly, the healthcare industry is a vast industry consisting of various kinds of data and large amounts. The rapid increase in the intelligent health industry involves extensive multimedia communication. The usage of IoMT has enhanced technological diversity in e-healthcare. In an IoMT-driven network, healthcare acts as one of the most sensitive applications. The data generated in this domain are highly confidential and cannot be shared or accessed by any ordinary person.

Nowadays, people are becoming increasingly concerned about their health as technology advances. As a result of this, many healthcare facilities have sprouted up. According to a study conducted by the United States Census Bureau [1], the elderly population in the United States will rapidly increase from 2010 to 2030. In Japan, the number of people in need of care is estimated to be between 140 and 160 million [2]. The Taiwan External Trade Development Council hosted the Taiwan International Senior Lifestyle and Health Care Show in June 2014 [3]. In the present days, the national debate is becoming increasingly focused on health care. As a result, several studies have merged technological goods into the healthcare business, which has become a norm.

The usage of wireless sensor networks in health care has increased significantly in recent years [4, 5]. The cloud is the preferable solution for medical sensor data processing and storage not only because of the cost benefits but also because of the scalability and flexibility needs. Though deploying medical sensor networks on the cloud may be more technically advantageous, processing sensitive personal data is an excellent concern for raising various security and privacy issues.

Encryption and access control are significant topics in the cloud world related to it. Different researchers have offered other access control systems. The DAC, MAC, RBAC, and IBE are some of the several access control techniques that may be used to ensure that data and sharing are as safe as possible. Several researchers have used attribute-based encryption (ABE) [6] because of the limitations of such access control mechanisms. The framework for effective data exchange for the health system was presented by Balamurugan et al. [7]. They utilized a framework built on the notion of a community cloud, and data sharing is restricted. The study makes no mention of encryption mechanisms for cloud data storage. The framework's installation costs are comparatively significant to other cloud environments since they employ private cloud infrastructure built on their premises.

The layered architecture for the health systems utilizing the cloud [8] proposes a comprehensive structure for storing data in the cloud based on the multi-cloud idea described in the referenced paper. The research presents a three-way storage system based on data criticality and encryption through the RC5 scheme. Access control becomes a significant issue despite the structure appearing to be efficient.

Though the results appear favorable, data exchange is limited to the clusters involved. The system model is not widely used [9]. The author presented an effective approach for resolving cloud interoperability difficulties [10]. But still, the access control is a problem once again in this.

Sahai and Waters [11] proposed a method to get around IBE's limitations in their paper Cipher text-ABE. They have presented a collision-free technique for delegation of authority delegation. The method also employs a threshold mechanism and sophisticated creation and decryption phases. The notion of a single trusted authority is employed in the CP-ABE scheme. In order to understand fine-grained access control, hierarchical ABE was later implemented [12]. Even though providing an excellent solution to fine-grained access control, it is challenging to put into action.

Additionally, the compound characteristics are not supported. Melissa Chase reported multiauthority ABE. It was based on the assumption that the cloud server is untrustworthy and that there is always a significant danger of relying only on a single authority for receiving credentials.

There are a lot of difficulties with user attribute revocation and decryption in the existing model. To avoid the decoding cost of encrypted text, Green et al. recommend outsourcing the access control to some other server. Even though the approach appears to be economical as it saves cost and time, it has major data verifiability issues. To address these challenges, Junzuo Loi devised an algorithm for outsourced data verification [13]. Ming Li presents a data-sharing architecture. He provides an access control technique that is lightweight and used for essential cloud applications, which outlines the many forms of ABE accessible and how they correspond with the application. Younis et al. [14], in their research work, proposed a novel access control mechanism (AC3) useful in the case of cloud computing.

Martin Giese et al. [15] demonstrated many functions of big data to be used in IoT as well as several techniques for IoT product design. In this article, the author addresses issues that may arise in the data used in the healthcare industry. These issues may be classified as flexibility, scalability, and elasticity. The author proposes novel data extraction method from databases on a single platform. Talal H et al. [16] highlight several services that are highly common in nature. He also mentions that data search is quite challenging in healthcare. Many data are collected and stored on these servers. The bulk storage allows the user or organization to make judgments regarding their business development. In terms of data, it supports a study by providing better understanding of cloud and other fields. Wi-Fi or a network medium is used to send data from the sensors. Initially, the data is kept on the local devices. When it comes to data storage, it has its own set of constraints.

From April 2008 to February 2012, an EU framework project ProSafe (Promoting patient Safety and quality improvement in critical care) was developed [17]. In addition to foreign work with Brazil, the partnership included seven partners from different countries. The project's purpose was to give ICUs a way to assess their problems and strengths by comparing them to those of other ICUs and to find opportunities and trends for enhancing critical care. To achieve the aim, the ICU must compare its performance to that of others on a regular, easy, rigorous, and confident basis. As a result, the team determined what data required to be gathered from the ICU and how it should be delivered. The ICUs that took part in the study were given Case Report Forms to fill out. There were two different sorts of forms made. The first was a set of documents that detailed the ICU unit's services, facilities, medical personnel, capabilities, and the hospital where it was housed. Patients are tracked using the forms in the second group when they are admitted to and treated in a specific ICU unit. A multilanguage web application was designed to collect these forms online.

Vipul Goyal et al. [18] introduced ABE in 2006 for fine-grained encryption access control. KP-ABE method was used to encrypt the data. Each user has been allocated to a tree access structure utilizing fuzzy identification-based encryption to prevent collaboration among users. In this type of encryption, identity has descriptive characteristics. Fine-grained access control was achieved using KP-ABE and fuzzy identity-based encryption. Hierarchical identity-based encryption was used as the delegation approach in this study, which introduced a large amount of overhead. Bilinear Diffie-Hellman Assumption and Chosen Ciphertext Attack security models were used to assess the performance.

To increase privacy and security, Melissa Chase and Sherman S.M. Chow [19] proposed multiauthority-attribute-based encryption (MA-ABE) in 2009. The MA-ABE needs a lot of attribute authority for key management. One of the MA-ABE Schaech's flaws is that it is vulnerable to a collision-resistant attack from a centralized authority that is entrusted with distributing users' keys. For eliminating trusted authority, the approach Distributed Pseudo-random function was utilized. With MA-ABE, the anonymous key usage protocol was employed to improve user privacy. The Diffie Hellman exchange conducted security analysis and confidentiality. For security analysis, an attribute-based selection approach was applied. To assess efficiency, performance criteria such as tolerance DKG instance, attribute authentication key size, user key size, and ciphertext size are employed.

Poornima and T. Rajendran [20] proposed a hybrid encryption technique to increase HASBE security for cloud computing in 2014. The upgraded HASBE combines hierarchical attribute-set encryption with ciphertext-policy-attribute encryption. Compound features of several users were used in a hierarchical framework to enable fine-grained access control and flexibility.

In the recent years, different approaches have also been adopted to assure the security of EPR across the network [21]. Many researchers have researched this domain of reversible data hiding (RDH) and developed numerous methods to provide security and integrity while using IoT. These RDH methods were developed employing a variety of ideas, including interpolation, compression, data transfer, watermarking with tamper localization, and difference expansion. In the recent years, digital watermarking of medical data such as images/videos and EPR has been used as a strong tool to overcome the problems of RDH [22]. The following are a few watermarking-related research projects.

A watermarking strategy for medical photographs was proposed by N. Solanki and S. K. Malik [23]. The RSA (asymmetric cryptography) approach was used to encrypt the watermark, which was then placed into ROI using the DWT algorithm. However, before genuine embedding can take place, the recommended approach needs preprocessing. Parah et al. suggested a high-capacity RDH approach for medical photos in [24]. This method creates high-quality stego-images, but the capacity was limited (i.e., 0.75 bpp). Therefore it was reduced to 0.5 bpp while incorporating a watermark or maintaining tamper detection scope. S. Lee presented a transform domain-based reversible watermarking solution in [25]. For information embedding, this includes high-frequency coefficients. It creates higher-quality watermarked photos, but it comes at a hefty expense in terms of computation.

T. C. Lu introduced a unique reversible data concealing approach based on interpolation, modulus function, and re-encoding strategy in [26] in response to more discussion regarding RDH methodology. Although the approach has a big payload of roughly 3 bpp, it is imperceptible. A. K. Singh et al. proposed a major non-blind and undetectable dual watermarking strategy for telemedicine applications in [27] to increase the robustness and imperceptibility. His plan was to combine the DWT's spatial-frequency localization capabilities with the SVD's visual perception quality. However, the payload was once again a significant disadvantage. A reversible data concealing approach with excellent imperceptibility was also presented by Bhardwaj and Aggarwal (2018). This approach embeds n secret bits per block by breaking into n sub-blocks. But, it results in a high bit error rate.

According to the thorough review of the literature, most documented reversible data embedding technologies have payload constraints, high computation costs, and tamper detection capabilities. We aim to overcome the problem of reduced payload and greater computational cost in our technique by keeping reversibility (to help in proper diagnosis) and assuring security by inserting a fragile watermark in the spatial domain for tamper detection.

The key works in this domain are listed in Table 1 along with the major findings.

Conclusion

This chapter discussed the importance of the IoT and machine learning in healthcare. A brief history of the IoT is given and the present scenario is discussed. The chapter discussed various application areas in which IoT can be used in the healthcare domain. The key benefits of using IoT in the healthcare domain are discussed in detail. The various IoT devices available in the market for healthcare support are briefed in this chapter. A few related works in RDH for healthcare sector are briefly discussed, along with the advantages and disadvantages. This chapter will be a useful reference for researchers who work in the domain of IoT and machine learning methods in healthcare.

Table 1: Research Work and Key Findings.

Author – year	Title of the paper	Research methodology	Findings and conclusion
Bansal et al. (2015)	A hybrid data encryption mechanism using RSA and Blowfish.	They suggested a system for implementing a cryptography approach using a field-programmable gate array (FPGA).	Because of its low cost and strong security, it is extremely effective. However, it takes up a lot of the key's space (448 bits)
Akomolafe et al. (2017)	A novel framework for data storage using hybrid cryptographic mechanism	They used the AES, Blake2b, and Schnorr Signature algorithms to enable secure data storage in the suggested technique. Because data is encrypted on the client side before being uploaded to the cloud, the service provider is unaware of the personal encryption mechanism.	Multimedia data such as music, picture, and video are not supported by the suggested technique.
Aruna Malik, Geeta Sikka, Harsh K Verma (2020)	A reversible data hiding scheme for interpolated images based on pixel intensity range.	They suggested a new reversible data concealing strategy for upscaling the original picture and hiding hidden data in the upscaled/interpolated image as well as an unique interpolation.	It uses the features of the human visual system to adaptively integrate the secret data into the interpolated pixels, resulting in a stego-image with a high concealing capacity and good quality.
Priyank Khare, Vinay Kumar Srivastava (2020)	A secured and robust medical image watermarking approach for protecting integrity of medical images.	For telemedicine applications, a unique medical image watermarking (MIW) approach is presented. The watermarking approach is developed using the homomorphic transform (HT), redundant discrete wavelet transform (RDWT), and singular value decomposition (SVD) transformations.	The watermark image is protected against modification or unethical acts by a double layer of protection given by AT chaotic mapping.

(continued)

Table 1 (continued)

Author – year	Title of the paper	Research methodology	Findings and conclusion
Navdeep Singh et al. (2015)	A framework for encrypting user data before delivering to cloud through a pair of methods.	The AES approach is used to encrypt user data, while the RSA algorithm is used to encrypt the secret key. For decryption, the same procedure is used. Even DoS assaults on the cloud were mitigated using the hybrid method.	One disadvantage of this framework is that it takes longer to finish the full procedure than other methods.
Shabir A. Parah, Javaid. A. Kaw et al. (2020)	Efficient security and authentication for Edge-based Internet of medical things.	The suggested method is based on a combination of innovative left data mapping (LDM), pixel repetition method (PRM), RC4 encryption, and checksum computation. PRM is used to upscale an input picture of size MN, ensuring reversibility while reducing computing complexity.	The proposed effort exceeds the current state of the art and is expected to have a favorable influence on security issues in smart healthcare services and society.
Nazir A. Loana, Shabir A. Parah et al. (2018)	A reversible and secure patient information hiding system for IoT-driven e-health.	Optimal Pixel Repetition (OPR), which turns every pixel of the input picture to a 2×2 block to assist reversibility by guaranteeing all the pixels in a 2×2 block have distinct values, is a unique high-capacity and reversible data concealing solution for safely integrating EPR within medical images.	Both subjective and objective indicators have been used to evaluate the suggested framework. The imperceptibility and fragility study is completed, and the findings are presented and discussed in depth.
Bouchti et al. (2016)	A new hybrid architecture based on cryptographic as a service (CaaS)	It was used to provide cloud clients permission to utilize their cryptographic keys and activities. For the security of healthcare data, they employed the homomorphic algorithm and RSA.	The suggested plan was built on the open source and open stack platform. So, there might be some security concerns.

References

[1] U.S. Census Bureau web site, http://www.census.gov/. Access date: 11–06–2022.

[2] S. Dhiviya, S. Malathy, and D. R. Kumar, “Internet of Things (IoT) elements, trends and applications,” Journal of Computational and Theoretical Nanoscience, vol. 15, no. 5, pp. 1639–1643, 2018.

[3] Taiwan Int'l Senior Lifestyle and Health Care Show web site, http://www.sencare.com.tw/en_US/index.html. Access date: 12–06–2022.

[4] J. Ko, C. Lu, M. Srivastava, J. Stankovic, A. Terzis, and M. Welsh, “Wireless sensor networks for healthcare,” Proceedings of the IEEE, vol. 98, no. 11, pp. 1947–1960, 2010.

[5] J. Biswas, J. Maniyeri, K. Gopalakrishnan, L. Shue, P.Eugene, H. Palit, F. Y. Siang, L. L. Seng, and L. Xiaorong, “Processing of wearable sensor data on the cloud – A step towards scaling of continuous monitoring of health and wellbeing,” in Engineering in Medicine and Biology Society (EMBC), 2010 Annual International Conference of the IEEE, 2010, pp. 3860–3863.

[6] N. S. Kumar, G. R. Lakshmi, and B. Balamurugan, “Enhanced attribute based encryption for cloud computing,” Procedia Computer Science, vol. 46, pp. 689–696, 2015, Proceedings of the International Conference on Information and Communication Technologies, {ICICT} 2014, 3–5 December 2014 at Bolgatty Palace & Island Resort, Kochi, India. [Online]. Available: http://www.sciencedirect.com/science/article/pii/S187705091500191X. Access date: 12–06–2022.

[7] B. Balamurugan, P. Venkata Krishna, N. Kumar, and G. Rajyalakshmi, “An efficient framework for health system based on hybrid cloud with abe-outsourced decryption,” in Artificial Intelligence and Evolutionary Algorithms in Engineering Systems, Ser. Advances in Intelligent Systems and Computing, L. P. Suresh, S. S. Dash, and B. K. Panigrahi, Eds. Springer, India, 2015, vol. 325, pp. 41–49. [Online]. Available: http://dx.doi.org/10.1007/978-81-322-2135. Access date: 11–06–2022.

[8] B. Balamurugan, P. Venkata Krishna, L. Rajya, and N. Saravana Kumar, “Layered storage architecture for health system using cloud,” in 2014 International Conference on Advanced Communication Control and Computing Technologies (ICACCCT), May 2014, pp. 1795–1800.

[9] B. Balamurugan, P. Krishna, G. Rajya Lakshmi, and N. Kumar, “Cloud cluster communication for critical applications accessing c-mpich,” in 2014 International Conference on Embedded Systems (ICES), July 2014, pp. 145–150.

[10] B. Balamurugan, N. S. Kumar, G. V. R. Lakshmi, and R. N. S. Shanmuga, “Common cloud architecture for cloud interoperability,” in Proceedings of the 2014 International Conference on Information and Communication Technology for Competitive Strategies, ser. ICTCS '14. ACM, New York, NY, USA, 2014, pp. 10:1–10:6. [Online]. Available: http://doi.acm.org/10.1145/2677855.2677865. Access date: 10–06–2022.

[11] J. Bethencourt, A. Sahai, and B. Waters, “Ciphertext-policy attributebased encryption,” in 2007 IEEE Symposium on Security and Privacy, SP '07, May 2007, pp. 321–334.

[12] Z. Wan, J. Liu, and R. Deng, “Hasbe: A hierarchical attribute-based solution for flexible and scalable access control in cloud computing,” IEEE Transactions on Information Forensics and Security, vol. 7, no. 2, pp. 743–754, April 2012.

[13] J. Lai, R. Deng, C. Guan, and J. Weng, “Attribute-based encryption with verifiable outsourced decryption,” IEEE Transactions on Information Forensics and Security, vol. 8, no. 8, pp. 1343–1354, August 2013.

[14] Y. A. Younis, K. Kifayat, and M. Merabti, “An access control model for cloud computing,” Journal of Information Security and Applications, vol. 19, no. no. 1, pp. 45–60, 2014. [Online]. Available: http://www.sciencedirect.com/science/article/pii/S2214212614000222. Access date: 11–06–2022.

[15] M. Giese, A. Soylu, G. Vega-Gorgojo, A. Waaler, P. Haase, E. Jimenez-Ruiz, D. Lanti, M. Rezk, G. Xiao, O. Ozcep, and R. Rosati, "Optique: Zooming in on big data," vol. 48, no. 03, March 2015.
[16] T. H. Noor, Q. Z. Sheng, A. H. H. Ngu, and S. Dustdar, "Analysis of web-scale cloud services," vol. 18, no. 4, July, August 2014.
[17] T. Prasanth, M. Gunasekaran, and D. R. Kumar, "Big data applications on health care," in 2018 4th International Conference on Computing Communication and Automation (ICCCA), December 2018.
[18] V. Goyal, O. Pandey, A. Sahai, and B. Waters, "Attributebased encryption for fine-grained access control of encrypted data," in Proceedings of the 13th ACM conference on Computer and Communications Security, October 2006, pp. 89–98.
[19] M. Chase and S. S. Chow, "Improving privacy and security in multi-authority attribute-based encryption," in Proceedings of the 16th ACM conference on Computer and Communications Security, November 2009, pp. 121–130.
[20] B. Poornima and T. Rajendran, "Improving cloud security by enhanced HASBE using hybrid encryption scheme," in 2014 World Congress on Computing and Communication Technologies. IEEE, 2013, pp. 312–314.
[21] S. Gull, S. A. Parah, and K. Muhammad, "Reversible data hiding exploiting Huffman encoding with dual images for IoMT based healthcare," Computer Communications, vol. 163, pp. 134–149, 2020.
[22] S. A. Parah, J. A. Sheikh, F. Ahad, N. A. Loan, and G. M. Bhat, "Information hiding in medical images: A robust medical image watermarking system for E-healthcare," Multimedia Tools and Applications, vol. 76, no. 8, pp. 10599–10633, 2017.
[23] N. Solanki and S. K. Malik, "ROI based medical image watermarking with zero distortion and enhanced security," International Journal of Education and Computer Science, vol. 10, pp. 40–48, 2014.
[24] F. Ahad, S. A. Parah, J. A. Sheikh, and G. M. Bhat, "Hiding clinical information in medical images: A new high capacity and reversible data hiding technique," Journal of Biomedical Informatics, vol. 66, pp. 214–230, 2017.
[25] S. Lee, C. D. Yoo, and T. O. N. Kalker, "Reversible image watermarking based on integer-to-integer wavelet transform," IEEE Transactions on Information Forensics and Security, vol. 2, pp. 321–330, 2007.
[26] T. C. Lu, "Interpolation-based hiding scheme using the modulus function and re-encoding strategy," Signal Processing, vol. 142, pp. 244–259, 2018.
[27] A. K. Singh, B. Kumar, M. Dave, and A. Mohan, "Robust and imperceptible dual watermarking for telemedicine application," Wireless Personal Communications, vol. 80, pp. 1415–1433, 2015.

Reethika A, Jagadesh T, M S Kanivarshini

Nutrition Food Recognition Using Deep Learning Algorithm for Physically Challenged Human Being

Abstract: Deep learning just surpassed the advancement in the process of food image identification. Nutrient, a novel deep learning architecture and a classification solution of image pixel classification of healthy or unhealthy food, is described in this research. This work is mainly focused for physically challenged people of their food consumption. Some people cannot be able to know whether the food is good to their health or not; hence, we proposed a method to recognize the food in a proper way. The new framework was made to approximate nutrition of the food at the image level using food-pics dataset that contains nutrient images. Convolutional neural networks, a deep learning approach that has been used successfully in image recognition and classification tasks, have been trained with nutrition image training data, and a high classification success value has been achieved. The arithmetic average nutritional values of the five highest predicted nutritional results obtained from the net were taken as the approximate nutrient content. These help to recognize the nutrition of food by the physically challenged people by themselves without any help of others.

Keywords: Food consumption, nutrition, image recognition, physically challenged

Intoduction

The need for automatic food recognition for food and drink appreciation classifications has developed in place of individuals who convert more conscious value of physically challenged people. Classifications provide automatic recognition sustenance and drink items, but they can evaluate their alimentary values and assemble them alimentary valuation and scheduling, which can assist patients with various dietary boundaries as complete as healthy people preventing nutrition-allied diseases.

Foods are often malleable items, making it challenging to determine their condition. Moreover, food categories have a great intraclass (meals gaze extremely different), but stumpy interclass discrepancy, identifying food even more difficult. A

Reethika A, Department of Electronics and Communication Engineering, KPR Institute of Engineering and Technology, Coimbatore, India, e-mail: reethikaanandan@gmail.com
Jagadesh T, Department of Electronics and Communication Engineering, KPR Institute of Engineering and Technology, Coimbatore, India
M S Kanivarshini, HCL Technologies, 71/A, Chokaramman nagar, Erode, Tamil Nadu, India, e-mail: kanivardhinis@hcl.com

https://doi.org/10.1515/9783110750584-007

problem with drinking acknowledgment is photographs of drink items, which can only provide a limited amount of information; for example, color of the drink and the density of the juice are examples of such information. Based on the constraints, both food and beverage images can identify a tough processer via a vision-based task.

We used deep neural networks (NNs) to solve this problem [1]. Several difficulties in computer vision need human defining of complex features, which is difficult and time-consuming. Deep learning solves this problem by allowing computer models with numerous handling layers to attributes and use them to incoming data. Deep learning algorithms greatly improve the number of research fields, including computer vision [1]. We are employing convolutional neural networks (CNNs) in particular, where the sort of NN responds to covering the area of optical field [2].

The CNN is very well adapted to computer vision, and the objective of these systems is comparable to that of vision systems, which is to interpret input images. Image is served into a CNN, and it undergoes sequence processes as input passes through the network layers. The training is then carried out by adjusting the layer settings in each iteration. CNNs have three types of layers: convolutional, fully connected, and pooling layers.

Filters were included in convolutional layers and are trained to recognize indisputable structures in input data filters as illustrated in Figure 1. Fully connected layers combine output data from other layers in order to extract higher order information as mentioned in figure 2. The pooling layers downsample data because they do not take parameters, but are not normally counted toward the overall NN layer depth in figure 3.

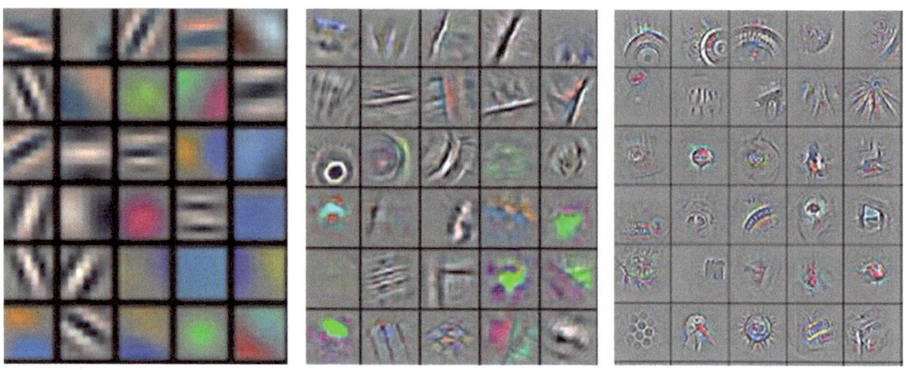

Figure 1: The layer of neural network was used to learn example filters, with simple characteristics such as edge and frequency orientation being represented.

This chapter is structured as follows: subject of food image detection and recognition is discussed in Introduction; image datasets and acquisition are explained in methods and detailed NutriNet and other CNN models that were tried as well as

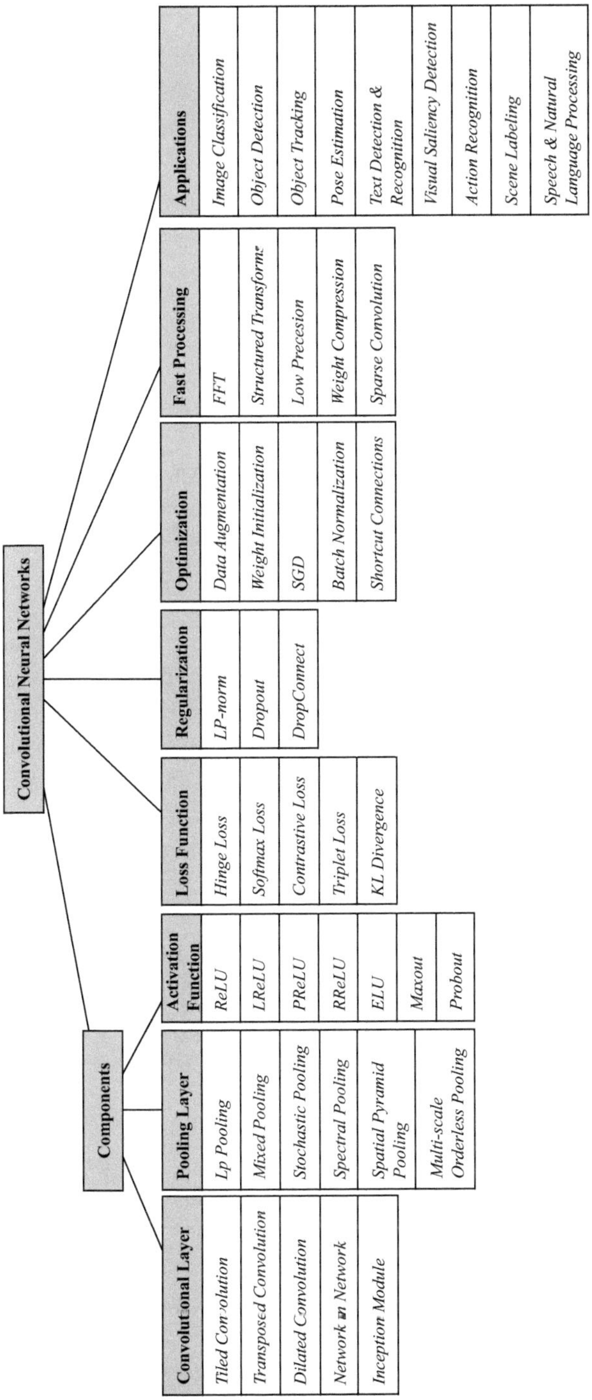

Figure 2: Hierarchically structured taxonomy of this survey.

their preparation course; next Segment is connected with exercise constituent that has been employed; The deep NN models' training and testing results; NutriNet architecture's testing results comprising an image dataset that created determination. The study comprises work argument portion, and finishes the article by providing a summary of the effort completed thus far as healthy as potential future exertion.

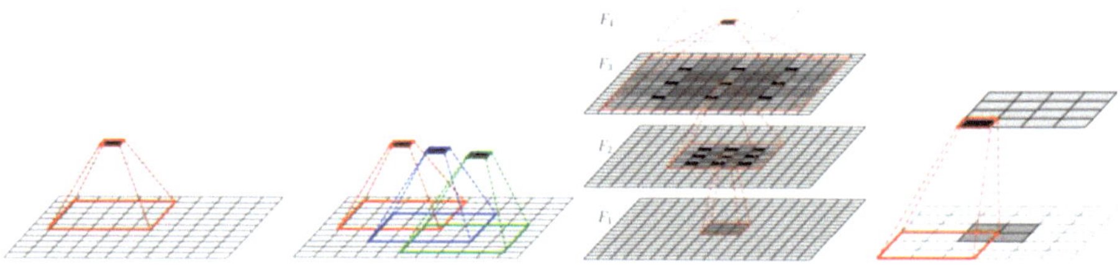

Figure 3: Convolution, tiled convolution, dilated convolution, and deconvolution.

Materials and Methods

Image datasets for food and drink are essentially aimed for nutrient that makes physically challenged people to have a confident in picking up food since it allows for over-all topographies, to avoid over fitting or overlapping while using the model, which describes the random noise rather than familiarity in algorithm as Machine learning. As a result, the objective was to create a dataset containing products based on food, which is denoted as several photos make feasible, by including the photos of meals and beverages based on their concerned country region, that would help the dietary assessment algorithm to perform better in its final application.

Because food and beverage kinds differ from location to location, a localized picture collection provides a more realistic depiction of meals besides beverages that would be recognized in practice. The data homework and model exercise procedure are not limited to images of specified country, so class labels as well as in other languages could be provided.

Recipe images shall be obtained from websites to create the image recognition dataset. This seemed suitable given the vast number of users on popular recipe websites, the majority of whom share not just the recipes but also the photographs of the finished dish. This strategy, however, had two major flaws: First, rather than particular dishes or drinks, the recipes simply listed the food groups (e.g., "meat dishes" and "vegetable dishes"). Because around where no designation guidelines or meals, the terms stood occasionally varied for the equal dish in addition to quite parallel to other sorts of serving dishes on the websites for individual recipes.

This stated that the registration result would be a wide-ranging category of food and beverage products, making recognition difficult. Furthermore, the model's utility is debatable because the findings would be so broad. Second describes the resultant image with a tiny dataset to a first-rate models comprised less than 10,0000 images. The justification for using only that food classifications differ significantly from making a unified recipe picture dataset difficult.

As a consequence, we created a recognition dataset of 60,000 food and drink picture programmers, each with 5,000 photos. The nature of web imaging searches, particularly of the consequences comprised inappropriate and low-quality photos. The preordained photographs like that have to be deleted in command to enhance the complete dataset eminence. This was accomplished by utilizing a deep CNN models that refers towards an identifying image as one or the double function which function which contains something different.

Figure 4: Various food identification.

On the recognition dataset, the perfect food and drink picture was employed, and photos tagged with "other" were deleted. As a result of the disparity in the amount of such photos between classes, the classes were imbalanced. The remaining photos have been rotated for four variants per image, and the intensification processes of three further figures were done by recognition dataset to expand the size and acquire more invariance, just as they did on the detection dataset. Nutrients of 19 photographs were flipped horizontally, given random color noise, and skyrocketed, 34% of the image boundaries were eliminated for every image as shown in figure 4.

There were seven different dissimilarities of each photo. AlexNet is the shortest of the three DNN designs, with five and three fully linked layers. AlexNet acquires fewer thorough settings since it is shallower, but it does it quickly. GoogLeNet has 22 layers, which is somewhat more than AlexNet. We employed ResNet-2 from architecture for this study, which has 256 layers and far subterranean than the other two ILSVRC-winning designs. Regardless of the modification in layer depth, AlexNet takes around the matching amount of constraints as ResNet, around 80 million, but GoogLeNet allows just about 4 million.

Some of the AlexNet, GoogLeNet, and ResNet do not employ fully connected layers because levels provide highest share of limitations, and these two designs may have a considerably greater number of layers without increasing the number of parameters. Dropout is a strategy for preventing overfitting in NN models by haphazardly eliminating units in the NN, as well as their connections shown in figure 5, throughout the process in all three of them.

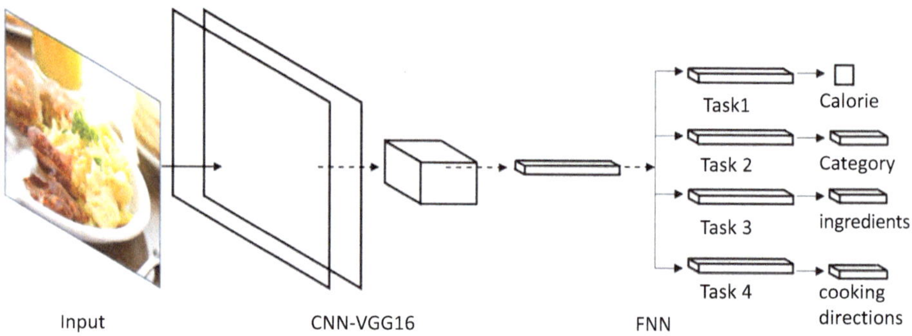

Figure 5: Processing stage of input.

Food access refers to the ability of consumers (individuals and households) to acquire food for a suitable and tolerable diet; the body building capacity is energetic and is based on the volume and form of domiciliary properties (personal health being one), organizational rules (e.g., imposed on those with frailties), and food scenes available. Food safekeeping is definite as admission by all persons at more times to sufficient food for a lively, fit lifetime, which embraces at the least:

(a) "The ready obtainability of nutritionally satisfactory and safe foods and (b) the secure ability to purchase adequate foods in as a friend sufficient ways, such as wanting resorting to violence."

This is because the levels provide greatest share of restrictions, these two designs may have a far greater number of layers without significantly increasing the number of parameters. Dropout is a strategy for preventing overfitting in NN by randomly eliminating items and their acquaintances during the preparation process.

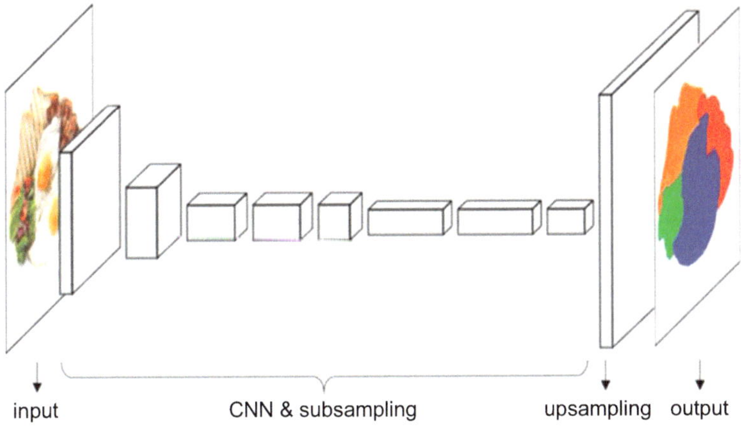

Figure 6: Proposed CNN-based assertiveness for physically challenged human-based food image recognition.

The network takes a 64 × 64 grayscale picture as input and generates an article mapped into two fully connected layers after many coatings of convolution and subsampling. Following the calculation of the fully connected layer, a 100-class output is obtained, representing the digital 0 to 9. Figure 6 shows the main mechanisms of network (CNN) of three (C1, C3, and C5), subsampling layers (S2, S4), and completely linked layers (F6 and output layer). A receptive field (also known as a fixed-size square or kernel) is chosen for the layer to divide convolution using equal scope square in the effort.

To progress the specified function of all pixel creative picture or article map is protected, the output feature is generated accordingly. To lower the dimension and prevent duplicate calculation, a subsampling procedure is performed surrounded by the article map after the convolution process. Finally, the fully linked layers are employed as a classifier to multidimensional feature maps into a fixed-size classification.

When utilizing character sample pictures to train, all of these deposits devor trainable limitations (weights) modified. Researchers are investing more effort into increasing the capabilities of describing visual information using more complicated models, according to some recent studies. With the large-scale tagged picture dataset ImageNet, a seven-layer model, comprising five CNN layers and two fully connected layers, is employed in the study. Subsequently more effort has been intensified of the number of layers and size although addressing the concerns of overfitting and saturation using Dropout and ReLU. ZFNet, VGG, and GoogLeNet were evolved in the following years with more complicated neurons, computing units, and layer topologies.

Experimental Results

Arrangement Architecture

The project goal is not to individually recognize the food on the salver, but likewise to cut the nutritional value of the meal. The system also assists users in controlling their food consumption. If a user's nutrient intake beats the required alimental allotment, the coordination will alert them. The suggested scheme's process is portrayed in Figure 4.

The scheme is made up of five main phases that are all linked together. These are the phases:

Task 1: Gathering food images
Task 2: Food item identification
Task 3: Using the food API to detect nutrients in the food
Task 4: Putting the fuzzy logic notion into practice
Task 5: Using an Android application to display the results

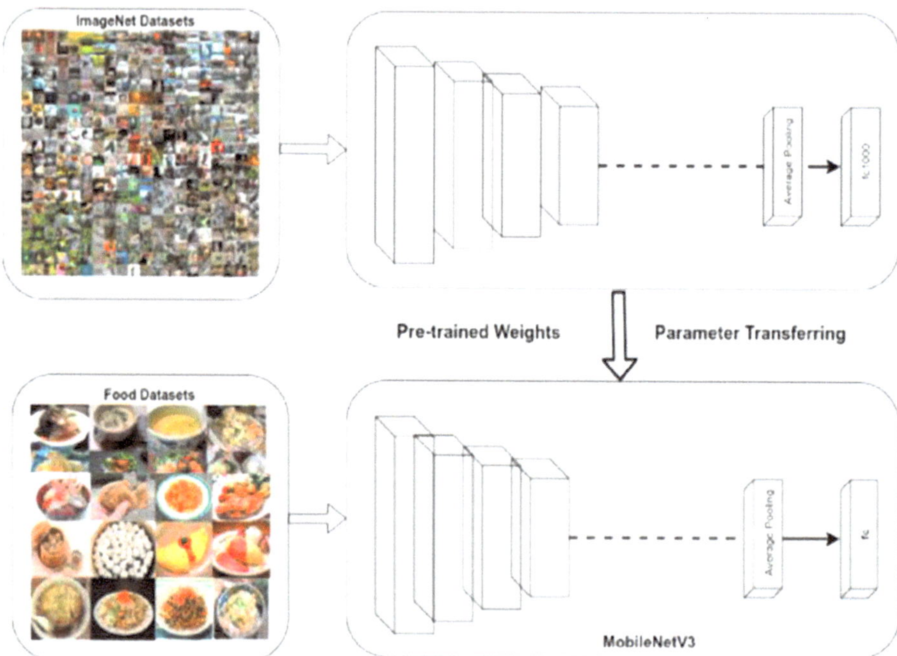

Figure 7: Parametric classification.

Capturing image data is the first stage in gathering data from various sources in order to build a machine learning model and understanding are known through image acquisition. We utilized the ETHZ Food-101 dataset in this study. Food-related photos that were mislabeled have been deleted from the collection. Picture augmentation is one of the strategy that may be utilized to add changes in the image, consequentially in an increased capacity of copies to generalize new imagr new images after training CNN on a large dataset.

By producing a modified version of pictures, image data augmentation affectedly increases the quantity of physical activity data. The Image Data Generator class in the Keras deep learning NN toolkit allows you to fit models using picture data augmentation. The machine learning model is built using Keras when the picture augmentation is completed in figure 8. A pretrained GoogLeNet/Inception model was employed in this study. Once the model has recognized the food category/food type, the image was sent as an input to the food nutrition API, which extracts the nourishment content of the food and provides it to the fuzzy logic system.

The nutritional information is compared with the required dietary allotment using the fuzzy logic algorithm. If individual nutrient, such as calorie, exceeds the recommended dietary allowance, the user will receive a caution letter in his or her mobile application to reduce nutrition intake, or else, the user's diet value will be displayed in an Person application designed for this purpose.

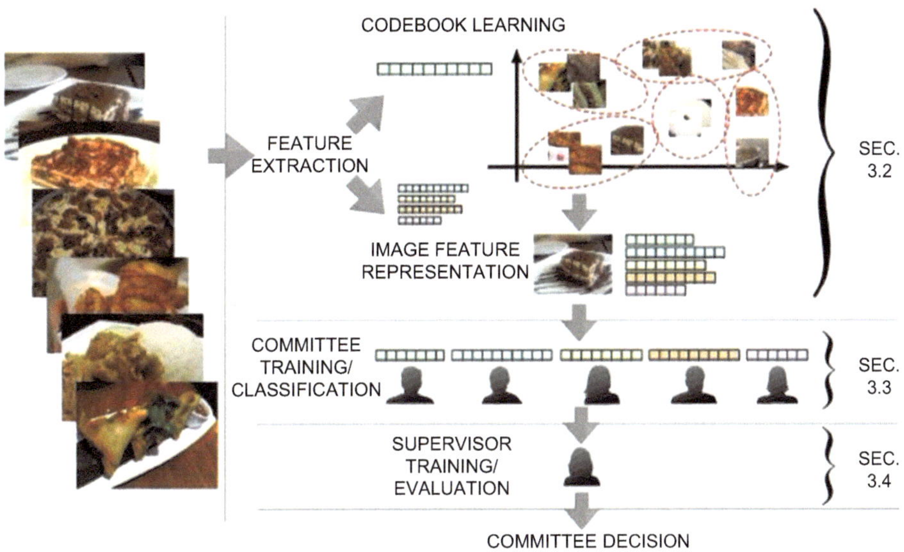

Figure 8: Multiple number of images for the food recognition.

Traditional Deep Neural Network

The traditional network architecture has certain limitations: when we train further into the simple network, training error lowers at first before dramatically rising. Overfitting is likely increasing as the network grows deeper:

- The number of increase in restriction leads to increased computational resources and expense.
- It takes a very big quantity of data to train CNN. For domain-specific difficulties, researchers do not always have adequate datasets.
- CNNs take anything from a week to a month to train.

The critical to build a model that can overaw the constraints of prevailing models after examining this section. One method is to build a sparsely connected network, which entails snowballing the number of layers of networks and number of elements at every tier.

1	Chicken Udon
2	Lettuce, cabbage, tomato
3	Rice with clams
4	Pork miso soup
5	Boiled pork meat
6	Baked atka mackerel

Figure 9: Differentiation of food.

Image processing techniques are used to transfer learning as shown in figure 9. Instead of starting from the beginning, transfer learning employs a pattern-based approach to learning. Patterns were taught when tackling a separate problem in the pattern-based method. In a time-consuming approach, an accurate model may be developed via transfer learning.

In computer vision issues, the preskilled model can be used to prompt assignment knowledge. These models were developed using a big collection of benchmark data. We used such models to apply the remaining machine learning model's expertise to a new assignment.

This research uses Keras' pretrained inception model, commonly known as GoogLeNet. With a top eight error rate of 8.9%, the inception model, Google researchers came up with this model. Earlier models, such as AlexNet and VGC, used a basic convolution and max pooling stack module. Typically, creating a CNN, the type of filter to be used is determined based on the application. A heap of 60 layers of convolutions and a novel element starts by including in this model shown in figure 10. The module comprises max-pooling in the inception layer and concurrent calculation of convolutions of various sizes (5 × 5, 9 × 9, and 7 × 7 convolutional layer).

Figure 10: Normal human being's food identification.

Our research yielded two main results: the lately defined NutriNet architecture image recognition dataset, which covers an abundant greater number of dissimilar food sorts than earlier pains in the ground and also includes a widespread change of drinks. Additionally, all the photographs were identified through online, other researchers may reproduce it and even adjust it to local food and drink products.

Formerly suggested approaches for food recognition have low arrangement accurateness because of intricacy pictures, while drink image recognition methods were hitherto imaginary. Deep learning comes into play difficult-to-define features; hence, it is a better option. The findings of our study effort back up this assertion. Overfitting remains a concern with deep learning, even utilizing the dropout strategy, and picking the perfect at the point of working-out phase when justification harm stops falling.

Overfitting couldbe the reasons accuracy of real images makes lowery arranged from subcategory, along through further noise and shutting off pictures and statistic which can be used to recognize the dataset, it may contain some extraneous images, the dataset has been generated with a food and drink image detection model precision of 96.57%, which means common to all images mentioned as C shown in figure 11 and 12. As a result, the categorization precision for real-world photos may

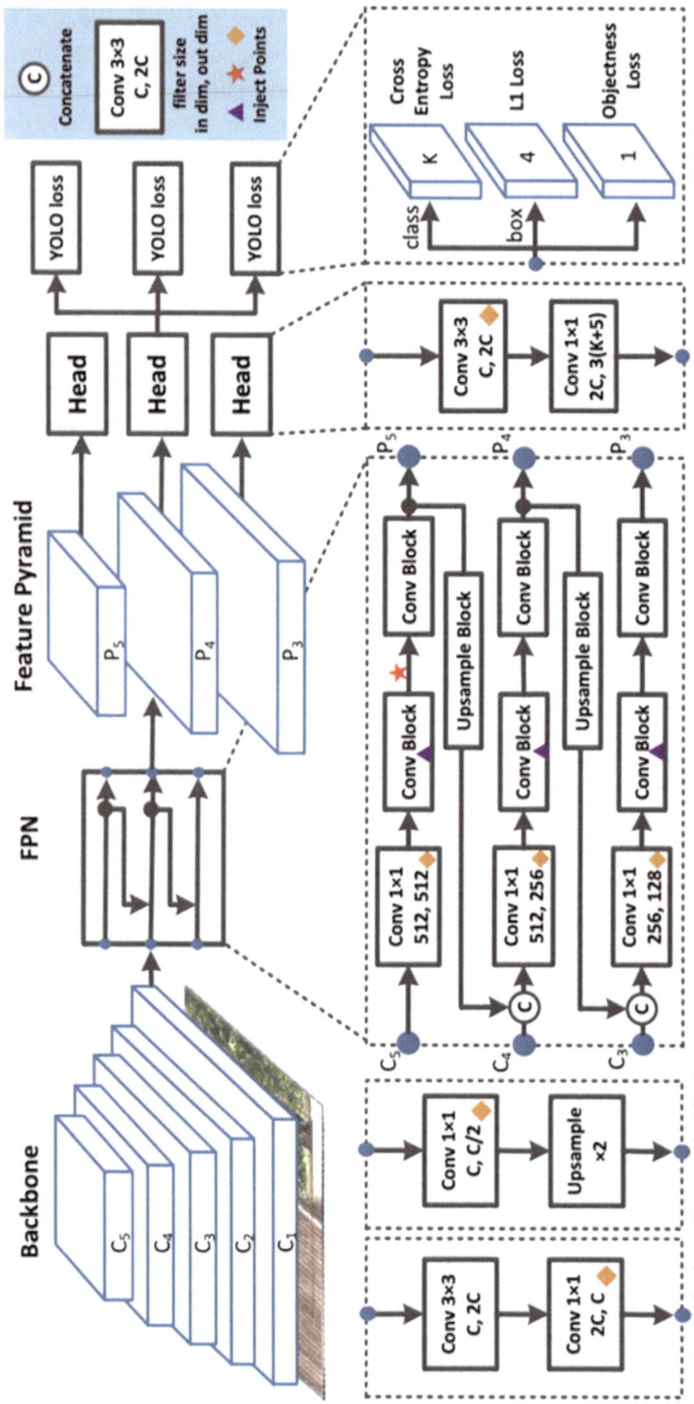

Figure 11: CNN model layer-based process.

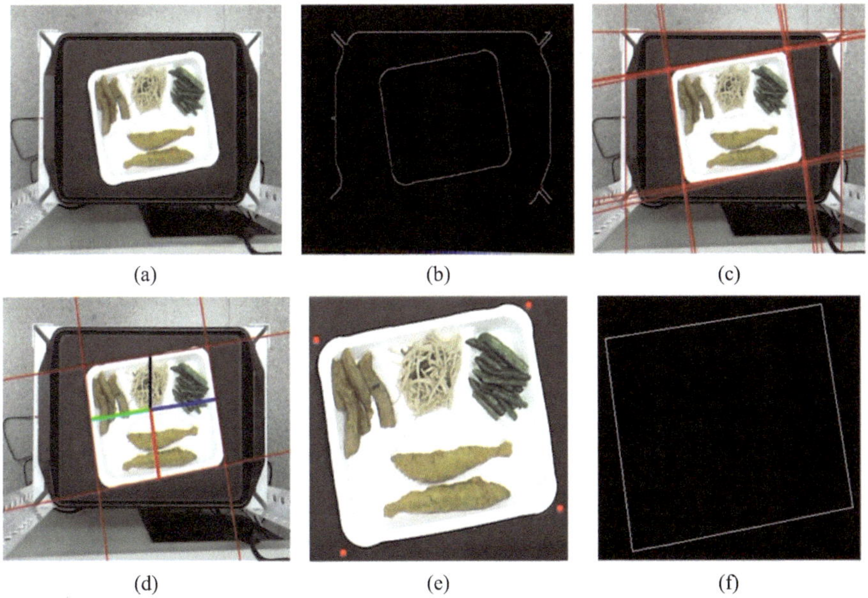

Figure 12: Multiple foods in a single platter identification.

suffer. Lastly, because of picture segmentation, the recognition job is made more difficult by the presence of irrelevant things in the images.

Conclusion

The present image detection and recognition organization of the foods that we developed as well as the NutriNet CNN architecture that we developed provide a complex cataloging accurateness. Images are collected in Google, keeping the model time from little to allow nearer modification. Our identification method is included into the Nutrition dietary assessment tool and it also includes refreshing classical with new photos in programs automatically.

The cecond phase of the study consists of architecture, which worked admirably and it might be better, particularly on everyday photos with noise and obstacles. We will use optimization methods to systematize phase as well because there are so many options for changing the design. We intend to further solve the problem of overfitting when more photographs are mechanically supplied to the dataset. A food and drink detection model might well be skilled to recognize photos with numerous food or drink components.

This model's outputs would each signify a different food or drink item, which might be fed into the current recognition model. An added solution to this challenge in combining the detection and recognition phases into a deep convolutional network is that more number of fault testing must be necessary for evaluation By using these methods, you may eventually achieve better results at lower prices, which will improve picture identification.

Using a CNN, we developed a technique for recognizing the existing split food items. The categorization was implemented patch by patch, and the period of both food items was determined using a vote process. We were able to extract enough examples to train a six-layer CNN using the patch-wise model and data increase technique. The system's efficacy was demonstrated by the trial findings, which showed an overall correctness of 84.9%. The consequences reported here are open, and the future study should embrace a more in-depth analysis of the best design as well as the network's training settings. Furthermore, using alternative classifiers with CNN features might improve performance even further.

References

[1] G. H. Beaton, J. Milner, P. Corey, V. McGuire, M. Cousins, E. Stewart et al., "Sources of variance in 24-hour dietary recall data: Implications for nutrition study design and interpretation," American Journal of Clinical Nutrition (USA), 2008, vol.8, pp. 48–55.

[2] W. C. Willett, L. Sampson, M. J. Stampfer, B. Rosner, C. Bain, J. Witschi et al., "Reproducibility and validity of a semiquantitative food frequency questionnaire," American Journal of Epidemiology, vol. 122, pp. 51–65, 2010.

[3] M. Buzzard, "24-hour dietary recall and food record methods," Monographs in Epidemiology and Biostatistics, vol. 1, pp. 50–73, 2018.

[4] K. Poslusna, J. Ruprich, J. H. de Vries, M. Jakubikova, and P. van't Veer, "Misreporting of energy and micronutrient intake estimated by food records and 24 hour recalls, control and adjustment methods in practice," British Journal of Nutrition, vol. 101, pp. S73–S85, 2019.

[5] R. Steele, "An overview of the state of the art of automated capture of dietary intake information," Critical Reviews in Food Science and Nutrition, 2018 vol.203, pp. 56–71.

[6] S. Yang, M. Chen, D. Pomerleau, and R. Sukthankar, "Food recognition using statistics of pairwise local features," in 2010 IEEE Conference on Computer Vision and Pattern Recognition (CVPR), 2018, pp. 2249–2256.

[7] Y. Matsuda and K. Yanai, "Multiple-food recognition considering co-occurrence employing manifold ranking," in 2012 21st International Conference on Pattern Recognition (ICPR), 2012, pp. 2017–2020.

[8] F. Zhu, M. Bosch, I. Woo, S. Kim, C. J. Boushey, D. S. Ebert et al., "The use of mobile devices in aiding dietary assessment and evaluation," IEEE Journal of Selected Topics in Signal Processing, vol. 4, pp. 756–766, 2018.

[9] B. L. Daugherty, T. E. Schap, R. Ettienne-Gittens, F. M. Zhu, M. Bosch, E. J. Delp et al., "Novel technologies for assessing dietary intake: Evaluating the usability of a mobile telephone food record among adults and adolescents," Journal of Medical Internet Research, vol. 14, 2017. pp 632–648.

[10] C. Xu, Y. He, N. Khannan, A. Parra, C. Boushey, and E. Delp, "Image-based food volume estimation," in Proceedings of the 5th international workshop on Multimedia for cooking & eating activities, 2018, pp. 75–80.
[11] TADA: Technology Assisted Dietary Assessment at Purdue University, West Lafayette, Indiana, USA, available at http://www.tadaproject.org/.
[12] C. K. Martin, T. Nicklas, B. Gunturk, J. B. Correa, H. R. Allen, and C. Champagne, "Measuring food intake with digital photography," Journal of Human Nutrition and Dietetics: The Official Journal of the British Dietetic Association, vol. 27, Suppl 1, pp. 72–81, January 2014.
[13] MyFitnessPal.com: Free Calorie Counter, Diet & Exercise Tracker, available at http://www.my fitnesspal.com/.
[14] MyNetDiary: The easiest and smartest free calorie counter and free food diary for iPhone, iPad, Android, and BlackBerry applications, available at http://www.mynetdiary.com/
[15] FatSecret: All Things Food and Diet, available at http://www.fatsecret.com/
[16] Y. Bengio, "Learning deep architectures for AI," Foundations and Trends® in Machine Learning, vol. 2, pp. 1–127, 2009.
[17] Meal Snap: Magical Meal Logging for iPhone, available at
[18] J. Deng, W. Dong, R. Socher, L.-J. Li, K. Li, and L. Fei-Fei, "ImageNet: A largescale hierarchical image database," in Proceedings of the IEEE Computer Vision and Pattern Recognition (CVPR), Miami, Florida, USA, 2015.
[19] T.-Y. Lin, M. Maire, S. Belongie, J. Hays, P. Perona, D. Ramanan et al., "Microsoft coco: Common objects in context," in Computer Vision–ECCV 2014 ed, Finland: Springer, 2014, pp. 740–755.
[20] A. Krizhevsky, I. Sutskever, and G. E. Hinton, "ImageNet classification with deep convolutional neural networks," in NIPS, 2017, p. 4.
[21] G. E. Hinton, S. Osindero, and Y.-W. Teh, "A fast learning algorithm for deep belief nets," Neural Computation, vol. 18, pp. 1527–1554, 2008.
[22] Scientists See Promise in Deep-Learning Programs, by John Markoff, New York Times. Available: http://www.nytimes.com/2012/11/24/science/scientists-seeadvances-in-deep-learning-a-part-of-artificial-intelligence.html
[23] Y. Bengio, L. Yao, G. Alain, and P. Vincent, "Generalized denoising auto-encoders as generative models," in Advances in Neural Information Processing Systems, 2013, pp. 899–907.
[24] R. Salakhutdinov and G. E. Hinton, "Deep boltzmann machines," in International Conference on Artificial Intelligence and Statistics, 2007, pp. 448–455.
[25] R. Girshick, J. Donahue, T. Darrell, and J. Malik, "Rich feature hierarchies for accurate object detection and semantic segmentation," in 2014 IEEE Conference on Computer Vision and Pattern Recognition (CVPR), 2014, pp. 580–587.
[26] K. Simonyan and A. Zisserman, "Very deep convolutional networks for large-scale image recognition," Computer Science- Neural and Evolutionary Computing, https://doi.org/10.48550/arXiv.1312.4400 Focus to learn more arXiv preprint arXiv:1409.1556, 2014.
[27] M. D. Zeiler and R. Fergus, "Visualizing and understanding convolutional networks," in Computer vision–ECCV 2019 ed Germany,: Springer, 2014, pp. 818–833.
[28] C. Szegedy, W. Liu, Y. Jia, P. Sermanet, S. Reed, D. Anguelov et al., "Going deeper with convolutions," in Proceedings of the IEEE Conference on Computer Vision and Pattern Recognition, 2020, pp. 1–9.
[29] V. Nair and G. E. Hinton, "Rectified linear units improve restricted boltzmann machines," in Proceedings of the 27th International Conference on Machine Learning (ICML-10), 2018, pp. 807–814.

[30] C. Szegedy, W. Liu, Y. Jia, P. Sermanet, S. Reed, D. Anguelov et al., "Going deeper with convolutions," in 2015 IEEE Conference on Computer Vision and Pattern cognition (CVPR), 2019.
[31] Y. Kawano and K. Yanai, "Foodcam: A real-time food recognition system on a smartphone," Multimedia Tools and Applications, vol. 74, pp. 5263–5287, 2018.
[32] M. Chen, K. Dhingra, W. Wu, L. Yang, and R. Sukthankar, "PFID: Pittsburgh fastfood image dataset," in 2009 16th IEEE International Conference on Image Processing (ICIP), 2019, pp. 289–292.
[33] Y. Kawano and K. Yanai, "Food image recognition with deep convolutional features," in Proceedings of the 2014 ACM International Joint Conference on Pervasive and Ubiquitous Computing: Adjunct Publication, 2018, pp. 589–593.
[34] Y. LeCun, L. Bottou, Y. Bengio, and P. Haffner, "Gradient-based learning aplied to document recognition," Proceedings of the IEEE vol. 86, pp. 2278–2324, 2008.
[35] D. H. Hubel and T. N. Wiesel, "Receptive fields, binocular interaction and functional architecture in the cat's visual cortex," The Journal of Physiology, vol. 160, pp. 106–154, 2010.
[36] M. Lin, Q. Chen, and S. Yan, "Network in network," Computer Science- Neural and Evolutionary Computing, https://doi.org/10.48550/arXiv.1312.4400 Focus to learn more arXiv preprint arXiv:1312.4400, 2013.
[37] K. He and J. Sun, "Convolutional neural networks at constrained time cost," in Proceedings of the IEEE Conference on Computer Vision and Pattern Recognition, 2015, pp. 5353–5360.
[38] Y. Jia, E. Shelhamer, J. Donahue, S. Karayev, J. Long, R. Girshick et al., "Caffe: Convolutional architecture for fast feature embedding," in Proceedings of the ACM International Conference on Multimedia, 2014, pp. 675–678.
[39] L. Bossard, M. Guillaumin, and L. Van Gool, "Food-101–mining discriminative components with random forests," in Computer Vision–ECCV 2014 ed, Switzerland: Springer, 2014, pp. 446–461.

K. Renuka Devi, S. Suganyadevi, K. Balasamy

Healthcare Data Analysis Using Deep Learning Paradigm

Abstract: In the present decades, analysis of healthcare domain plays a significant role for research purposes and it depends more on computer technology. The analysis of medical data is one of the most primary factors for healthcare domain that has to be done by the machine learning (ML) intelligent systems. Normally, the healthcare data is composed of huge volume of sensitive data such as patient's information, health insurance information, and patient's medical histories. The handling and analyzing those data for extracting important facts is one of the tedious process. To resolve this issue, ML plays a vital role in analyzing the above gathered data. ML utilizes several statistical methods as well as complex algorithms to process the healthcare data to predict the results. For efficient data analysis, those methods are categorized as supervised, unsupervised, and semi-supervised. To analyze those sensitive data in a deeper manner, the neural network under ML approach promotes a significant role. In-specific, deep learning (DL) models are utilized for pattern recognition in the healthcare domain. DL is a component of the artificial intelligence and machine learning (AI/ML) paradigm that mimics how people learn. So, this technique is highly recommended for analyzing and recognizing patterns in clinical data. This chapter aims to provide detailed review of various DL approaches, DL algorithms, different DL applications, and its utilization in healthcare, followed by its issues and challenges.

Keywords: Deep learning, deep neural networks, healthcare domain, machine learning, medical data

Introduction

Machine learning (ML) model is among the most rapidly growing research area that has been utilized in different applications like web and text mining, detection of spam, recommender systems, and information retrieval. While considering various kinds of ML algorithms, deep learning (DL) is one of the most prominent areas

K. Renuka Devi, Department of Information Technology, Dr. Mahalingam College of Engineering and Technology, Pollachi, India
S. Suganyadevi, Department of ECE, KPR Institute of Engineering and Technology, Coimbatore, India
K. Balasamy, Department of AI and DS, Bannari amman Institute of Technology, Sathyamangalam, Erode, India

https://doi.org/10.1515/9783110750584-008

which is utilized in the above-said applications [1]. DL is also represented as deep-structured learning which is considered as broader area of ML methods based on ideas from artificial neural networks. The foundation of DL is found in the convolutional neural network (CNN). However, it performs better than those techniques. DL utilizes transformation and graph technologies for constructing multilayer learning models. It has been utilized in recent applications like video processing, audio processing, and natural language processing (NLP).

The accuracy of the input-data representation is essential for the effectiveness of a ML algorithm. The input data is the important factor for deciding the performance of ML algorithm [2]. So the appropriate data representation will have high impact than imperfect data utilization. The main aim of DL algorithm is to attain feature extraction which is accomplished by means of automatic mechanism. Those algorithms consist of multilayer learning models where the foremost layer extracts the features of less importance and the end layer extracts the features with higher importance.

DL is highly recommended:

- For solving complex problems to discover hidden patterns from data.
- To understand the relationship between various kinds of interdependent variables.

The Concept of Deep Structured Learning

DL is a type of AI and ML concepts where it performs exactly the same way as human individuals acquire specific type of cognitive knowledge [3]. DL is considered to be the most vital model for data science since it includes different techniques like predictive and statistical modeling. In the aspect of data scientists, the concept of DL is extremely important who are handling large quantity of data for performing different operations like data collection, data analyzation, and data interpretation [4]. Hence, the DL tends to make this process much faster and easier to execute.

Predictive analytics is the technique for making predictions about future outcome results by utilizing various historical data that has been combined with data mining, ML, and statistical modeling techniques [5]. So, the DL model is highly recommended for making the predictive analytics in a more automated way. When we compare DL algorithms with traditional ML algorithms, those DL algorithms tend to possess hierarchy of abstraction and complexity of algorithm execution.

The DL algorithm works in a manner that it applies non-linear transformations to the input and also utilizes those transformations for creating statistical model as output [6]. The process of iterations continues until the result has achieved the level of accuracy. The keyword "deep" in the term DL denotes the number of processing layers through which data pass to reach that accuracy. The relationship of AI, ML, and DL has been depicted in Figure 1.

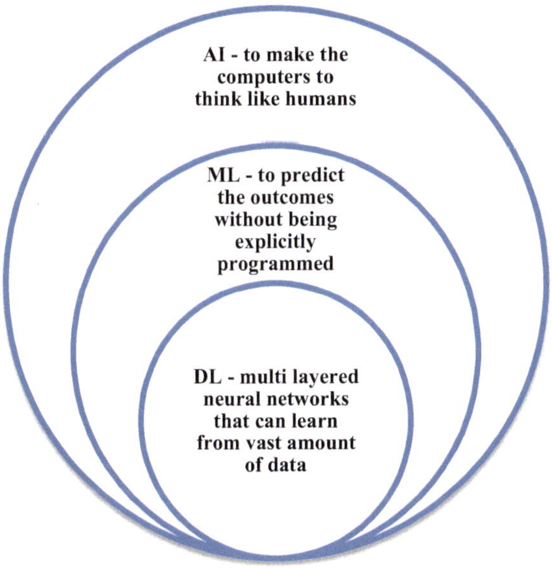

Figure 1: AI, ML, and DL – the relationship.

A class of ML models known as "deep learning" imitates the structure of the human brain. The DL algorithms result in various conclusions which is very similar to humans by analyzing logical data [7]. DL does this through the use of neural networks, a multilayered framework of algorithms. It has three layers: the input layer, the concealed layer, and the output layer. The input layer accepts an input like "x," and n-hidden layers accept the input and analyze it in each layer of structure to achieve more accurate result and the resultant has been moved onto the output layer as $\vec{y}$. The above process is depicted in Figure 2.

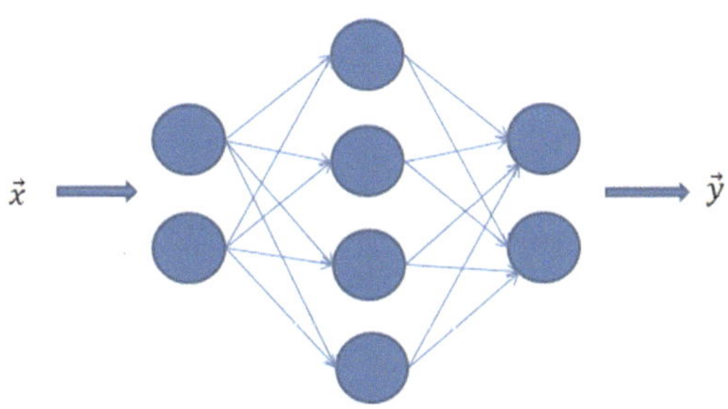

Figure 2: Neural network – concept.

In general, neural networks are used to carry out different operations such as clustering, classification, and regression. Unlabeled data can be grouped or sorted based on similarity relationships among the samples [8]. For classification, it is required to train the network for dataset which is labeled. This training is needed for classifying the samples into various categories. Neural networks behave in the same way and also carry out the same functions as traditional algorithms of ML [9, 10].

Benefits of Deep Learning

- Feature extraction is one of the advantages of DL where those features will be traced automatically to fetch the appropriate outcome.
- Time-saving process.
- The same neural network model can be utilized for different applications.
- The multiple legitimate issues can be overcome using the DL architecture.

Classification of DL Approach

DL is used to analyze huge amount of data to get the desired results and so the DL methods are classified as supervised, semi-supervised, and hybrid techniques [11, 12]. Under supervised learning strategy, it consists of CNN algorithm. Under unsupervised learning strategy, it consists of recurrent neural network (RNN) algorithm [13]. Under semi-supervised learning strategy, it consists of generative adversarial networks (GAN) algorithm [14]. The classification of DL approaches is depicted in Figure 3.

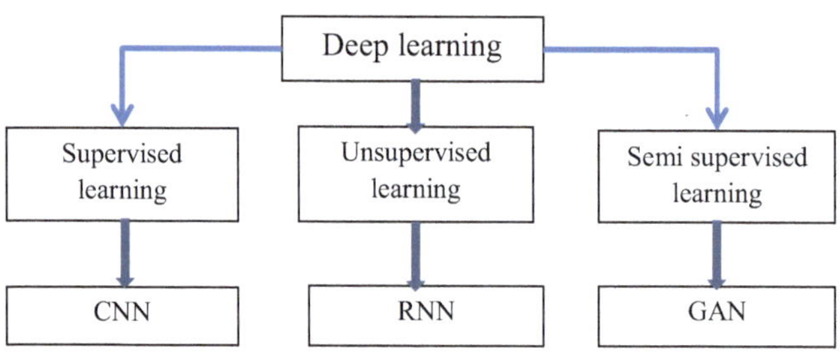

Figure 3: Classification of DL approach.

Supervised Deep Learning Strategy

The supervised DL strategy focused on labeled data. When dealing with SDLS, this technique has a collection of inputs and the resulting outputs [15]. The parameters of network have been updated in order to gain the preferred outputs based on the input parameters. When the training has been in a right direction, this strategy will result in a right solution based on the input attributes [16]. Under supervised DL strategy, a CNN plays a significant role.

Benefit

- SDLS can be able to collect or generate data based on previous knowledge.

Drawback

- The calculation of decision boundary might be a tedious process if the values are not inside the original class values.

Convolutional Neural Network (CNN or ConvNet)

In the area of DL, the ConvNet algorithm is well-known and extensively used. CNN has been widely employed in a variety of applications including voice processing and facial recognition. Neurons in human and livestock brains provided the inspiration for CNNs, and their structure is comparable to that of a traditional neural network [17]. In simple, ConvNet, which takes input as the image, then incorporates various objects in the input image and analyzes them to distinguish each image in order to retrieve the better output. The step of preprocessing is not much required in CNN as when contrasted to other methods for classification [18].

An input layer, a convolutional layer, a fully connected layer, and an output layer are some of the layers that make up the CNN, as shown in Figure 4.

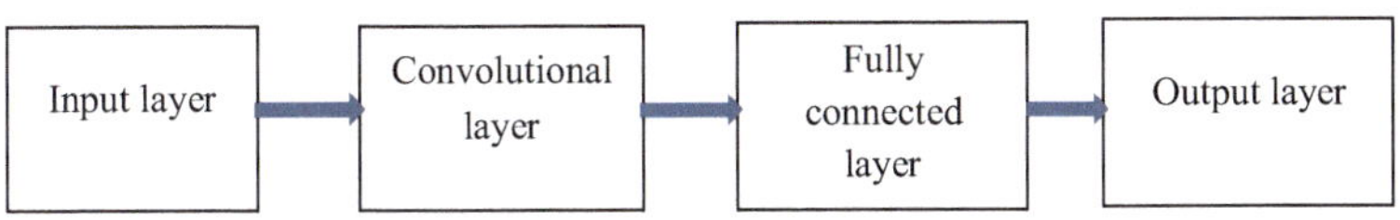

Figure 4: CNN process.

Input layer: The data has been studied through the input layer.
Convolutional layer: This layer structure is used to retrieve various qualities or features from the input source.
Fully connected layer: The output of the convolutional layer was sent as an input to the fully connected layer in order to get the output.
Output layer: This layer is called as fully generated layer because it has $(n+1)m$ parameters, where "n" denotes the quantity of inputs and "m" denotes the quantity of outputs.

The output for CNN can be calculated using the following formula:

$$O = \frac{n - f + 2p}{s} + 1 \tag{1}$$

where n stands for n number of inputs, f represents the filter, p denotes padding, and s denotes stride.

Benefits

- CNN produces results with high accuracy in image recognition.
- It has the capacity to detect the vital features automatically in an autonomic manner.

Drawbacks

- It will not encode the data and object.
- It requires huge amount of training data to produce accurate results.

Unsupervised Deep Learning Strategy

Unsupervised deep learning strategy tends to analyze the features in a dataset with no labels and it has an autonomic execution. This method allows the learning process to be carried out even when there is no labeled data available [19]. So, the agent tends to learn the significant traits or inner representations needed to uncover the input data's unidentified structure or relationships. Several attributes of DL method had performed better for both clustering and dimensionality reduction technique [20]. RNN comes under this category which also includes GRUs and LSTMs, and these techniques have been utilized in a wide range of applications.

Benefits

- For producing better and effective processing models.
- To unlock previously unsolvable problems.

Drawbacks

- It requires human intervention to process and understand the features and patterns than other methods.
- It is not sure that the generated results might be useful all the time.

Recurrent Neural Network

RNNs are one of the most potent varieties of neural networks and efficient one. It is one of the most potential algorithms now in use as it is the only one with an internal memory [21]. Due to this internal memory, RNNs have the ability to remember critical details about the input they receive, allowing them in determining the next with greater accuracy. This technique has been used extensively for sequential data, including speech, text, and time series, and RNN analyzes the data in much deeper manner and its sequence has been efficiently executed as it is compared to other such algorithms [22]. In a nutshell, RNNs are a subset of neural network algorithms that are primarily concerned with sequential data. It operates in a manner identical to that of the human mind. RNN has the capacity of remembering its past attributes. Since it contains two distinct inputs: the present and the past, it can be able to remember the past values as most of the algorithms do not, which is depicted in Figure 5.

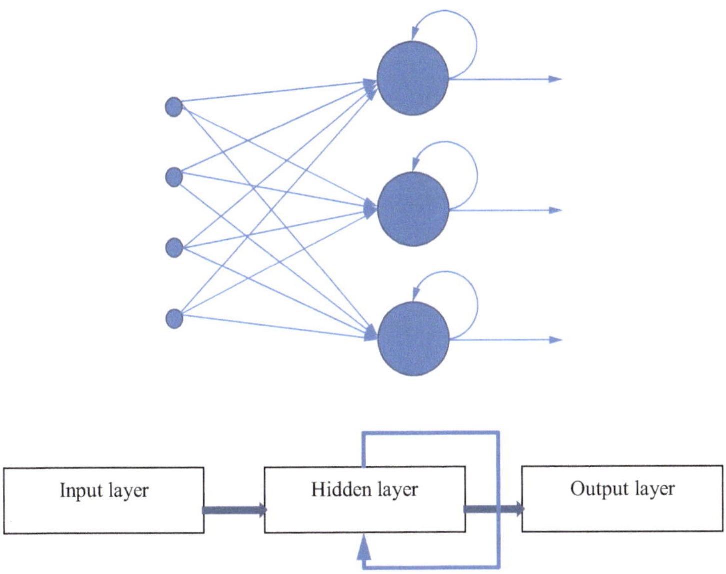

Figure 5: RNN process.

RNN can be calculated using the following formula:

$$y_t = wh_t \tag{2}$$

where y_t denotes the output, w represents weight at the output layer, and h_t denotes the activation function of RNN.

Benefits
- RNN can be able to model a record in large amount since the result of each pattern depends on the previous results.
- It can be used along with convolutional layers to extend the property of pixel features.

Drawbacks
- RNN has the drawback of gradient exploding and vanishing problem.
- Training RNN is one of the tedious process.

Semi-Supervised Deep Learning Strategy

Semi-supervised strategy of learning is a method which is used to train the model by utilizing an adequate amount of labeled data which is extracted from huge amount of unlabeled data [23]. In a nutshell, SSDLS is a kind of process which is based on unlabeled datasets where a small subset of dataset will contain the class labels.

Semi-supervised learning makes an effort to extrapolate from a limited number of labeled data points to efficiently categorize additional, previously unseen cases by internalizing concealed structures in the data. To efficiently handle the semi-supervised deep learning (SSDL), most of the labeled and unlabeled data should come from the same functional form of distribution [24].

Benefits
- The requirement of labeled data can be minimized.
- Cost for labeling the data can be minimized.

Drawbacks
- Not applicable for network level data.
- It has less accuracy.

Generative Adversarial Networks

GANs are considered as generative models of network which was designed by Ian Goodfellow and his colleagues in 2014. It consists of two kinds of neural networks such as generator and discriminator [25]. Those two networks depend on each other such that the loss of one network acts as the other network's gain. The generator feature aims to generate fake data which can be taken from the training data, and the job of a discriminator is to segregate data as authentic or fraudulent, which is generated from the generator which is depicted in Figure 6. The data can be audio, video, or image where the GANs can perform.

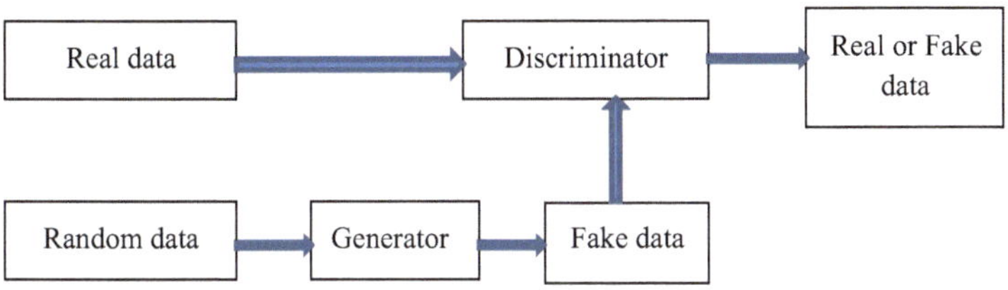

Figure 6: GAN – methodology.

In general, GANs are considered as game theory approach because the discriminator performs in the way not get deceived by the generator and generator performs in the way that it always try to deceive the discriminator [26]. Overall, both the kinds of network performs in their own way to prove its efficiency where output can be computed using the following equation:

$$V(D, G) = E_{x \sim p_{\text{data}}(x)}[\log D(x)] + E_{z \sim p_z(z)}[\log(1 - D(G(z))] \tag{3}$$

where G stands for generator, D represents discriminator, $p_{\text{data}}(x)$ indicates real-time data distribution, $P(z)$ denotes generator distribution, x represents sample data from $p_{\text{data}}(x)$, z represents sample data from $p(z)$, $D(x)$ represents network of discriminator, and $G(z)$ denotes network of generator.

Benefits

- GANs can produce data which is similar to the original one.
- It can be able to recognize different images and also the distance between them.

Drawbacks

- Harder to train the data – it requires various types of data to understand whether the model is working accurately or not.
- Generating text from speech is one of the tedious process.

Deep Learning in Healthcare Domain

In today's world, healthcare sector plays a significant role that holds huge amount of biomedical data and those data take up a crucial role in healthcare domain [27].

The goal of right healthcare is to provide the safe treatment to the right patients which should be offered at the right time. And this treatment should be based on different attributes such as patient's data, environment, health records of patients, and patients' location.

The availability of vast biomedical data attempts to bring different opportunities in healthcare domain. The association among those data in the dataset helps to identify and develop a different medical tool which is highly helpful for providing appropriate treatment to the patients [28, 29]. Generally, this domain requires an expert to create healthcare models to be utilized in ad hoc situation. The utilization of supervised and unsupervised models might miss some patterns during execution [30]. Hence the DL models enter into the healthcare domain to extract patterns by analyzing the data in more deeper manner such that it transforms the representational state from one level to the more abstract level. In tasks like speech recognition, natural language processing, and others, DL models excel.

The DL models in healthcare play a predominant role in extracting deep patterns by analyzing the healthcare records for providing the appropriate treatment to the patients. Due to the development of various methodological improvements, DL paradigms tend to promote new opportunities for enhancing the healthcare domain [31]. For instance, Google deep mind had planned to invent new technologies in healthcare.

There are several approaches in DL sector that can address the problems faced by medicinal field which is depicted in Figure 7. This includes higher evaluation of performance, end-to-end scheme for learning, and able to handle a large quantity of complex and multi-variety of data [32]. As a whole, the DL can also be able to overcome the several challenges such as availability of noisy data, analyzation of healthcare records to derive the patterns, and heterogeneous data [33].

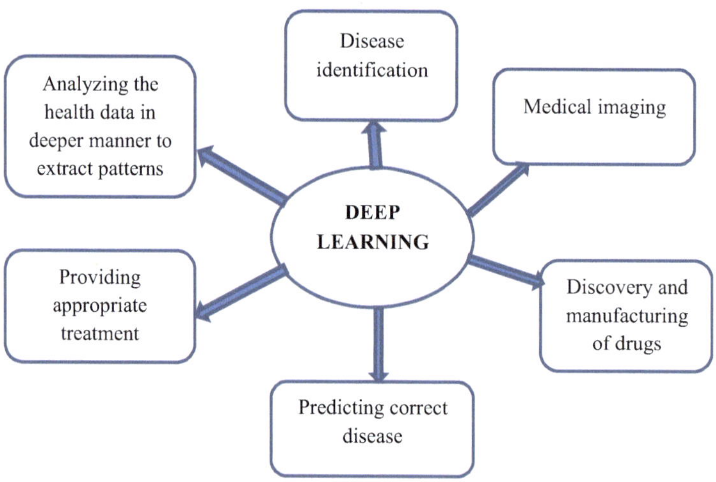

Figure 7: Deep learning in healthcare.

Challenges Faced by Healthcare Sector

Healthcare industry has been growing tremendously and the data handled in the domain has also been increasing at a higher rate. It significantly affects the progress of the country as well as the development and well-being of the society. But, this sector has been facing several challenges ranging from availability of data to the appropriate treatment to the patients [34]. There are several challenges related to healthcare sector which is depicted in Figure 8:

- The availability of healthcare systems and equipment.
- Tremendous growth of clinical data.
- Appropriate data for making better decisions.
- Legal challenges.
- Privacy and security issue.

The Availability of Healthcare Systems and Equipment

For providing necessary treatment to the patients, the clinical equipment should be available in the needed time. For effective and efficient service delivery, the availability and utilization of different healthcare devices have been focused at all points of health care system [35]. Due to the lack of such equipment and devices physically, the treatment may get delayed and leads to loss of lives.

Tremendous Growth of Clinical Data

The healthcare data has been increasing at a faster rate due to the handling of different variety of data for providing appropriate treatment to the patients [36]. Due to the growth of such data, the extraction of patterns also become difficult process and the handling of such data also become tedious. Because of this, the data become available anywhere and so the privacy issues have also been increasing [37].

Appropriate Data for Making Better Decisions

Because there is such a large amount of data available, choosing the right data for providing relevant treatment to the correct patients has been facing a serious issue [38]. When the wrong data has been selected and if it has been delivered to the patient, it will result in a serious issue. So, the suitable data selection and fetching accurate results have played a vital role in ensuring that appropriate treatment is provided to patients and also making better decisions [39].

Legal Challenges

The protection of data is one of the biggest concerns that every domain is facing in order to protect the information which is sensitive from the outside world [40]. When we deal with healthcare data, it contains a huge amount of sensitive data than any other domain, and so the strict laws have been implemented for protecting data compliance in medicinal business. This challenge could be overcome by an act called the Health Insurance Portability and Accountability Act for guaranteeing the security of medical records and medicinal practices for following suitable procedures [41].

Privacy and Security Issue

The privacy and security is the important factor that every domain should maintain for ensuring safe delivery of services to the customers. According to healthcare, this factor holds one step ahead than all the sectors [42]. It is because it was dealing with huge amount of sensitive data. The most common issue or threat the healthcare domain faces includes data breach, data theft, unauthorized access, and loss of data; hence, it should be handled with utmost care for delivering correct treatment to right patients at ease [43].

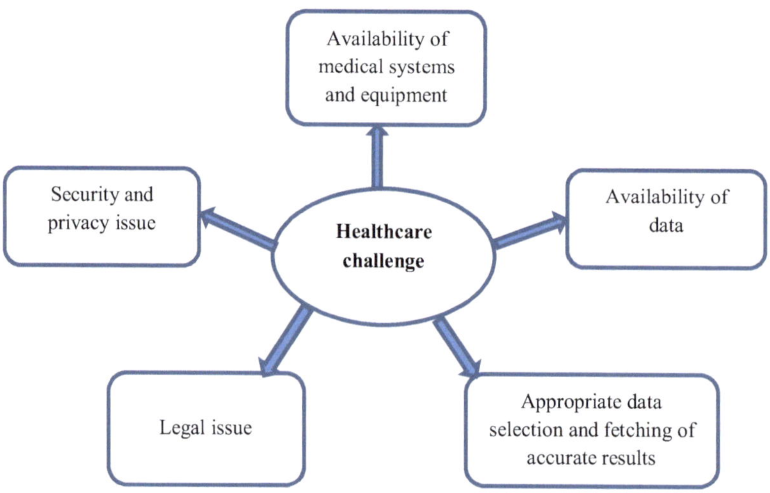

Figure 8: Challenges in healthcare.

Remedy by Deep Learning in Healthcare

DL is an area where it deals with huge amount of data to extract trends and patterns by analyzing those data in a deeper manner. When we deal with healthcare domain, one of the major challenges it was facing is that the dealing with huge amount of data and analyzing and retrieving the relevant data is one of the major concerns [44]. There comes the DL methodology where it fetches the data, analyzes them by passing through several layers, and fetches the results. This type of analyzation is very helpful for dealing with images in healthcare domain for matching and evaluating different healthcare images in order to provide appropriate treatment to the patients.

Another challenge faced by healthcare is that selection of relevant data by comparing with the previous results. The selection of those relevant data can be done efficiently by DL model [45]. The recurrent neural network model plays a significant role to overcome this challenge since this model works on the concept of remembering the previous data by connecting the present values. This model works efficiently for providing better results to the users.

Another issue is that healthcare domain facing is that retrieving accurate results for providing correct service to the patient. This can be overcome by GAN where the images in healthcare domain might consist of fake images; those results could be analyzed very deeply to differentiate the real and fake results [46]. GAN works in a significant manner for providing trustable results to the users.

Deep Learning – Applications

The DL implementation could be seen in various areas and the important one is discussed in this section. Those applications have been depicted in Figure 9.

Natural Language Processing

NLP is an integral part of AI which is utilized to understand how people share the information. In today's world, DL is used to attain a better performance on NLP tasks [47, 48]. Natural language processing is utilized to train machines through the concept of DL to fetch appropriate results to each kind of scenario. It provides the ability to machines not only to just read but also to determine and interpret the language of humans [49]. With the concept of NLP under DL model, machines could be able to identify the written or spoken text and also carry out tasks including speech recognition, text summarization, and sentiment analysis of the given data.

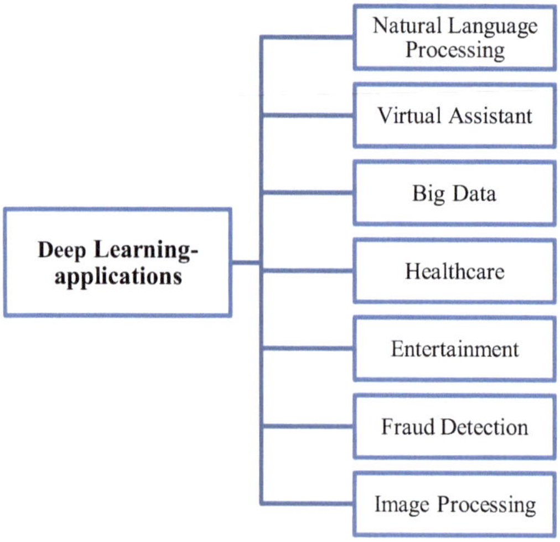

Figure 9: Deep learning applications.

Examples: Summarization of documents, classification of documents, twitter content analysis.

Virtual Assistant

Nowadays, the world becomes digitalized where the virtual assistants serve and play a significant part in the effort to close the gap between the digital and analogue worlds. Advanced DL, ML, and NLP techniques are used by a virtual assistant to extract difficult data, process them, and to extract the data more efficiently [50]. Those assistants could be able to provide general recommendations based on the users' profile, previous data, and location. Virtual assistants are largely utilized to automate various tasks like updation of system and transaction.

Examples: translation of speech to text, summarization of documents, organizing emails, carrying out various organizational activities.

Big Data

For carrying out efficient analysis of huge amount of unsupervised dataset, DL concepts play a significant role and used as a tool for big data analytics. DL's utilization of huge data is beneficial in extracting complicated patterns from massive amounts of data as well as data tagging, semantic indexing, fast data retrieval, and speeding discriminative tasks [51]. The algorithms under DL are used to improve the accuracy and performance during the real-time processing of data. The DL concepts are utilized to handle enormous amounts of unstructured data in order to automatically extract patterns and features.

Examples: Processing and extraction of complex data, classification, prediction, processing of unsupervised data.

Healthcare

DL tends to contribute major part for healthcare that diminishes the issue of health factors by cutting down costs. It mainly focused on the field of medicinal research [52]. DL has high potential for handling the unavoidable situation such as respiratory failures or in any risky conditions. It helps the healthcare sector to improvise the diagnostic and monitoring of patients [53]. For efficient result generation and providing appropriate treatment to the patients, DL is important in the clinical area since it can meet all of the users' requirements [54, 55].

Examples: Medical imaging, analysis of genome, drug discovery relevant to diseases, early prediction of diseases, etc.

Entertainment

Not only the DL can be able to provide appropriate solutions in the domain of healthcare, big data, and NLP, it can also be able to contribute to the entertainment world. There may be a situation where the platforms such as Amazon Prime and Netflix are providing recommendations or suggestions according to the user preferences; it was possible by DL. Such platforms can be able to generate insights based on performance-based and popularity-based [56]. Several OTT platforms are utilizing DL techniques to provide efficient recommendations to the users.

Examples: audio/video analysis, transcriptions, creation and editing of contents, analysis of audio and video, etc.

Fraud Detection

DL has a significant part in detecting frauds in the field of financial and banking sector. This becomes a bottleneck for the banking domain and also for the customers to make the online transactions efficient. DL has overcome this issue by incorporating pattern identification in transactions and credit scores in online [57]. This technique helps to prevent and detect the unusual transactions in the banking sector. Several ML strategies have been developed to overcome this problem but the DL might be helpful in minimizing the issue without any loss to the humans.

Examples: e-commerce transactions, online payment transactions, identity theft, etc.

Image Processing

DL highly contributes to the domain of image processing where it analyzes several images deeply to come to better representation of features without human intervention. It trains the data to fetch and provides appropriate results to the users. The DL network analyzes the images by passing those data across several layers and each layer fetches the data from the previous layer [58]. While passing the result to various layers, the data accuracy and learning details increase. This makes the user to get the appropriate results based on the requirements by utilizing DL capacity [59, 60].

Examples: classification of images, removal of noise from images, etc.

Conclusion

We focused on DL approaches in this study and how they play a predominant role in providing accurate results to the users. A brief review on different kinds of DL techniques such as supervised, unsupervised, and semi-supervised strategies has been discussed. Under those techniques, CNN, RNN, and GAN DL models have been focused and how it plays a major role in each domain has also been discussed. This paper also states how DL plays a huge part in healthcare domain to overcome several issues like the availability of healthcare systems and equipment, Tremendous growth of clinical data, appropriate data for making better decisions, legal challenges, and privacy and security issue by implementing the DL models. The integration of DL techniques in several applications like NLP, virtual assistant, big data, entertainment, healthcare, fraud detection, and image processing has also been discussed.

References

[1] R. N. Thomas and R. Gupta, “A survey on machine learning approaches and its techniques,” in 2020 IEEE International Students’ Conference on Electrical, Electronics and Computer Science (SCEECS), 2020, pp. 1–6, doi: 10.1109/SCEECS48394.2020.190.

[2] M. Wang, W. Fu, X. He, S. Hao, and X. Wu, “A survey on large-scale machine learning,” in IEEE Transactions on Knowledge and Data Engineering. doi: 10.1109/TKDE.2020.3015777.

[3] M. Dol and A. Geetha, “A learning transition from machine learning to deep learning: A survey,” in 2021 International Conference on Emerging Techniques in Computational Intelligence (ICETCI), 2021, pp. 89–94, doi: 10.1109/ICETCI51973.2021.9574066.

[4] S. Amiri, S. Salimzadeh, and A. S. Z. Belloum, "A survey of scalable deep learning frameworks," in 2019 15th International Conference on eScience (eScience), 2019, pp. 650–651, doi: 10.1109/eScience.2019.00102.
[5] H. Song, M. Kim, D. Park, Y. Shin, and J.-G. Lee, "Learning from noisy labels with deep neural networks: A survey," in IEEE Transactions on Neural Networks and Learning Systems, 2022. doi: 10.1109/TNNLS.2022.3152527.
[6] A. Yarali, "Applications of Artificial Intelligence, ML, and DL," in Intelligent Connectivity: AI, IoT, and 5G, IEEE, 2022, pp. 279–297. doi: 10.1002/9781119685265.ch16.
[7] C. Dalvi, M. Rathod, S. Patil, S. Gite, and K. Kotecha, "A survey of AI-Based facial emotion recognition: Features, ML & DL techniques, age-wise datasets and future directions," IEEE Access, vol. 9, pp. 165806–165840, 2021. doi: 10.1109/ACCESS.2021.3131733.
[8] M. Dol and A. Geetha, "A learning transition from machine learning to deep learning: A survey," in 2021 International Conference on Emerging Techniques in Computational Intelligence (ICETCI), 2021, pp. 89–94, doi: 10.1109/ICETCI51973.2021.9574066.
[9] M. I. Patel, S. Suthar, and J. Thakar, "Survey on image compression using machine learning and deep learning," in 2019 International Conference on Intelligent Computing and Control Systems (ICCS), 2019, pp. 1103–1105, doi: 10.1109/ICCS45141.2019.9065473.
[10] C. Fri and R. Elouahbi, "Machine Learning and Deep Learning applications in E-learning Systems: A literature survey using topic modeling approach," in 2020 6th IEEE Congress on Information Science and Technology (CiSt), 2020, pp. 267–273, doi: 10.1109/CiSt49399.2021.9357253.
[11] I. H. Sarker, "Deep learning: A comprehensive overview on techniques, taxonomy, applications and research directions," SN Computer Science, vol. 2, no. 420, 2021. doi: https://doi.org/10.1007/s42979-021-00815-1.
[12] K. R. Devi, G. Anandhi, and G. Saranya, "A systematic review of machine learning algorithms in data mining," in The Proceedings of International Workshop on Privacy Preserving Artificial Intelligence Techniques for Machine Learning, pp. 64–76, 2021.
[13] L. Alzubaidi, J. Zhang, A. J. Humaidi et al., "Review of deep learning: Concepts, CNN architectures, challenges, applications, future directions," Journal of Big Data, vol. 8, no. 53, 2021. doi: https://doi.org/10.1186/s40537-021-00444-8.
[14] A. Shrestha and A. Mahmood, "Review of deep learning algorithms and architectures," IEEE Access, 1–1, 2019. doi: 10.1109/ACCESS.2019.2912200.
[15] B. Fang, Y. Li, H. Zhang, and J. Chan, "Semi-supervised deep learning classification for hyperspectral image based on dual-strategy sample selection," Remote Sensing, vol. 10, no. 4, pp. 570–574, 2018. doi: https://doi.org/10.3390/rs10040574.
[16] R. Miorelli, A. Kulakovskyi, B. Chapuis, O. D'Almeida, and O. Mesnil, "Supervised learning strategy for classification and regression tasks applied to aeronautical structural health monitoring problems, " Ultrasonics, vol. 113, pp. 106372, 2021. doi: 10.1016/j.ultras.2021.106372.
[17] S. Albawi, T. A. Mohammed, and S. Al-Zawi, "Understanding of a convolutional neural network," in 2017 International Conference on Engineering and Technology (ICET), pp. 1–6, 2017, doi: 10.1109/ICEngTechnol.2017.8308186.
[18] J. Yue, W. Zhao, S. Mao, and H. Liu, "Spectral-spatial classification of hyperspectral images using deep convolutional neural networks," Remote Sensing Letters, vol. 6, pp. 468–477, 2015.
[19] M. Khanam, T. Mahboob, W. Imtiaz, H. Ghafoor, and R. Sehar, "A survey on unsupervised machine learning algorithms for automation, classification and maintenance," International Journal of Computer Applications, vol. 119, pp. 34–39, 2015. doi: 10.5120/21131-4058.

[20] Y. Chen, C. Yang, Y. Zhang, and Y. Li, "Conditional adaptation deep networks for unsupervised cross domain image classifcation," in 2019 14th IEEE Conference on Industrial Electronics and Applications (ICIEA), pp. 517–521, 2019, doi: 10.1109/ICIEA.2019.8834368.

[21] B. Chandra and R. K. Sharma, "On improving recurrent neural network for image classification," in 2017 International Joint Conference on Neural Networks (IJCNN), pp. 1904–1907, 2017, doi: 10.1109/IJCNN.2017.7966083.

[22] W. Yang, D. Zhang, and Y. Fu, "Research of a diagonal recurrent neural network and artificial neural networks," in 2016 International Symposium on Computer, Consumer and Control (IS3C), pp. 374–377, 2016, doi: 10.1109/IS3C.2016.103.

[23] Y. Reddy, V. Pulabaigari, and B. Eswara, "Semi-supervised learning: A brief review," International Journal of Engineering & Technology, vol. 7, no. 81, 2018. doi: 10.14419/ijet.v7i1.8.9977.

[24] S. Mandapati, S. Kadry, R. L. Kumar et al., "Deep learning model construction for a semi-supervised classification with feature learning," Complex & Intelligent Systems, 2022. doi: https://doi.org/10.1007/s40747-022-00641-9.

[25] A. Aggarwal, M. Mittal, and G. Battineni, "Generative adversarial network: An overview of theory and applications," International Journal of Information Management, 2021. doi: 10.1016/j.jjimei.2020.100004.

[26] J. Kim, S. Bu, and S. Cho, "Zero-day malware detection using transferred generative adversarial networks based on deep autoencoders," Information Science, 460–461, 2018. doi: https://doi.org/10.1016/j.ins.2018.04.092.

[27] L. Nie, M. Wang, L. Zhang, S. Yan, B. Zhang, and T.-S. Chua, "Disease inference from health-related questions via sparse deep learning," IEEE Transactions on Knowledge and Data Engineering, vol. 27, pp. 2107–2119, 2015. doi: https://doi.org/10.1109/TKDE.2015.2399298.

[28] K. Balasamy, N. Krishnaraj, J. Ramprasath, and P. Ramprakash, "A secure framework for protecting clinical data in medical IoT environment," Smart Healthcare System Design: Security and Privacy Aspects, 2021. doi: https://doi.org/10.1002/9781119792253.ch9.

[29] K. Balasamy, N. Krishnaraj, and K. Vijayalakshmi, "Improving the security of medical image through neuro-fuzzy based ROI selection for reliable transmission," Multimedia Tools and Applications, 2022. doi: https://doi.org/10.1007/s11042-022-12367-4.

[30] A. A. Abdullah, M. M. Hassan, and Y. T. Mustafa, "A review on bayesian deep learning in healthcare: Applications and challenges," IEEE Access. doi: 10.1109/ACCESS.2022.3163384.

[31] K. P. Subiksha, "Improvement in analyzing healthcare systems using deep learning architecture," in 2018 4th International Conference on Computing Communication and Automation (ICCCA), pp. 1–4, 2018, doi: 10.1109/CCAA.2018.8777545.

[32] L. Wang, X. Siyao, Y. Qian, and C. Huang, "A context-aware sensing strategy with deep reinforcement learning for smart healthcare," Pervasive and Mobile Computing, 2022. doi: https://doi.org/10.1016/j.pmcj.2022.101588.

[33] X. Zeng, S. Lin, and C. Liu, "Multi-view deep learning framework for predicting patient expenditure in healthcare," IEEE Open Journal of the Computer Society, vol. 2, pp. 62–71, 2021. doi: 10.1109/OJCS.2021.3052518.

[34] S. R. Sinha and B. Michele, "Health technologies and innovation in the global health arena," New England Journal of Medicine, vol. 9, 2011. doi: 10.1056/NEJMp1108040.

[35] R. Malkin and A. Keane, "Evidence-based approach to the maintenance of laboratory and medical equipment in resource-poor settings," Medical & Biological Engineering & Computing, vol. 48, pp. 721–726, 2010. doi: 10.1007/s11517-010-0630-1.

[36] L. Zhengwu and S. Jing, "Clinical data management: Current status, challenges, and future directions from industry perspectives," Open Access Journal of Clinical Trials, vol. 2, 2010. doi: 10.2147/OAJCT.S8172.

[37] G. Singh, D. Schulthess, N. Hughes, B. Vannieuwenhuyse, and D. Kalra, "Real world big data for clinical research and drug development," Drug Discovery Today, vol. 23, no. 3, pp. 652–660, 2018. doi: https://doi.org/10.1016/j.drudis.2017.12.002.
[38] M. Sousa, A. Pesqueira, C. Lemos, M. Sousa, and Á. Rocha, "Decision-making based on big data analytics for people management in healthcare organizations," Journal of Medical Systems, vol. 43, pp. 1–10, 2019. doi: 10.1007/s10916-019-1419-x.
[39] W. Raghupathi and V. Raghupathi, "Big data analytics in healthcare: Promise and potential," Health Information Science and Systems, vol. 2, no. 3, 2014. doi: https://doi.org/10.1186/2047-2501-2-3.
[40] S. Gerke, T. Minssen, and G. Cohen, "Ethical and legal challenges of artificial intelligence-driven healthcare," Artificial Intelligence in Healthcare, 295–336, 2020. doi: https://doi.org/10.1016/B978-0-12-818438-7.00012-5.
[41] C. M. Klugman, L. B. Dunn, J. Schwartz, and I. G. Cohen, "The ethics of smart pills and self-acting devices: Autonomy, truth-telling, and trust at the dawn of digital medicine," The American Journal of Bioethics, vol. 18, no. 9, pp. 38–47, 2018. doi: https://doi.org/10.1080/15265161.2018.1498933.
[42] K. Abouelmehdi, A. Beni-Hessane, and H. Khaloufi, "Big healthcare data: Preserving security and privacy," Journal of Big Data, vol. 5, no. 1, 2018. doi: https://doi.org/10.1186/s40537-017-0110-7.
[43] A. Appari and M. Johnson, "Information security and privacy in healthcare: Current state of research," International Journal of Internet and Enterprise Management, vol. 6, pp. 279–314, 2010. doi: 10.1504/IJIEM.2010.035624.
[44] C. M. Thwal, K. Thar, Y. L. Tun, and C. S. Hong, "Attention on personalized clinical decision support system: Federated learning approach," 2021 IEEE International Conference on Big Data and Smart Computing (BigComp), pp. 141–147, 2021. doi: 10.1109/BigComp51126.2021.00035.
[45] Jyotsna and P. Nand, "Enhancing Quality of service in IoT Healthcare services using Fog Computing," in 2021 2nd International Conference for Emerging Technology (INCET), pp. 1–5, 2021, doi: 10.1109/INCET51464.2021.9456341.
[46] Q. Zhang, B. Lian, P. Cao, Y. Sang, W. Huang, and L. Qi, "Multi-source medical data integration and mining for healthcare services," IEEE Access, vol. 8, pp. 165010–165017, 2020. doi: 10.1109/ACCESS.2020.3023332.
[47] S. Ramakrishnan, T. Gopalakrishnan, and K. Balasamy, "A wavelet based hybrid SVD algorithm for digital image watermarking," Signal & Image Processing An International Journal (SIPIJ), vol. 2, no. 3, pp. 157–174, 2011.
[48] T. Gopalakrishnan, S. Ramakrishnan, K. Balasamy, and A. S. M. Murugavel, "Semi fragile watermarking using Gaussian mixture model for malicious image attacks [C]," in 2011 World Congress on Information and Communication Technologies, pp. 120–125, 2011.
[49] S. K. A. Fahad and A. E. Yahya, "Inflectional review of deep learning on natural language processing," in 2018 International Conference on Smart Computing and Electronic Enterprise (ICSCEE), pp. 1–4, 2018, doi: 10.1109/ICSCEE.2018.8538416.
[50] G. Pirana, A. Sertbaş, and T. Ensari, "Sentence classification with deep learning method for virtual assistant applications," in 2019 3rd International Symposium on Multidisciplinary Studies and Innovative Technologies (ISMSIT), pp. 1–5, 2019, doi: 10.1109/ISMSIT.2019.8932888.
[51] J. Qiu, Q. Wu, G. Ding, Y. Xu, and S. Feng, "A survey of machine learning for big data processing," EURASIP Journal on Advances in Signal Processing, 2016. doi: 10.1186/s13634-016-0355-x.

[52] K. Balasamy and D. Shamia, "Feature extraction-based medical image watermarking using fuzzy-based median filter," IETE Journal of Research, 1–9, 2021. doi: https://doi.org/10.1080/03772063.2021.1893231.
[53] A. A. Abdullah, M. M. Hassan, and Y. T. Mustafa, "A review on bayesian deep learning in healthcare: Applications and challenges," IEEE Access. doi: 10.1109/ACCESS.2022.3163384.
[54] S. Suganyadevi, D. Shamia, and K. Balasamy, "An IoT-based diet monitoring healthcare system for women," Smart Healthcare System Design: Security and Privacy Aspects, 2021. doi: https://doi.org/10.1002/9781119792253.ch8.
[55] S. Suganyadevi, V. Seethalakshmi, and K. Balasamy, "A review on deep learning in medical image analysis," International Journal of Multimedia Information Retrieval, vol. 11, pp. 19–38, 2022.
[56] S. Suresh, N. Sinha, and M. Prusty, "Latent approach in entertainment industry using machine learning," International Research Journal on Advanced Science Hub, vol. 2, 2020. doi: 10.47392/irjash.2020.106.
[57] X. Kewei, B. Peng, Y. Jiang, and T. Lu, "A hybrid deep learning model for online fraud detection," in 2021 IEEE International Conference on Consumer Electronics and Computer Engineering (ICCECE), pp. 431–434, 2021, doi: 10.1109/ICCECE51280.2021.9342110.
[58] S. Karanam, Y. Srinivas, and M. Krishna, "Study on image processing using deep learning techniques," in Materials Today: Proceedings, 2020. doi: 10.1016/j.matpr.2020.09.536.
[59] K. Balasamy, N. Krishnaraj, and K. Vijayalakshmi, "An adaptive neuro-fuzzy based region selection and authenticating medical image through watermarking for secure communication," Wireless Personal Communications, vol. 122, pp. 2817–2837, 2022.
[60] K. Renuka Devi, J. Bhavithra, and Dr. A. Saradha, "Personalized nutrition recommendation for diabetic patients using improved K Means and Krill Herd optimization," International Journal of Scientific & Technology Research, vol. 9, no. 3, pp. 1076–1083, 2020.

R. Babu, Prithi Samuel, K. Jayashree

Cognitive Authentication for Smart Healthcare System

Abstract: For cognitive monitoring, that is, behavioral recognition and intellectual tracking of each individual, safe, and reliable monitoring is critical. In recent years, both industry and academia have become interested in cognitive authentication, often known as biometric recognition. The main purpose of cognitive authentication is to recognize, detect, and monitor dissimilarities in biological signatures, behavioral characteristics including facial gestures, biometrics, voices, eye gaze, mannerisms, and posture, among other things. Numerous developing biometric authentication systems are currently becoming incorporated and developed in multidisciplinary sectors such as personally identifiable information, accessibility, and asset monitoring systems. Whenever it comes to simple legalizing and identification surveillance difficulties, conventional security methods are not up to equivalence. To address these issues, an electroencephalogram-based cardiovascular technology is a promising choice for next-generation mechanisms due to brain signals reflecting/mimicking cognitive style and behaviors. Data encryption and identification tracking are included in today's modern advanced authentication systems. Users identify themselves by using words/passwords, yet that alone would be insufficient to adequately protect and secure private data. Because standard password protection lacks cognitive distinctiveness, it could be stolen and hacked by anybody. To improve a bad passcode and maintain it, the user must enhance the difficulty level and that will be extremely difficult to remember.

Keywords: Cognitive system, authentication, electroencephalogram, healthcare system

Introduction

In the world nowadays, Electronic-Medicare elucidations predominant in the exploring communal remains not only focus with solving enduring diseases. Mental fitness observation is one of the requirements that need attention from various researchers

R. Babu, Department of IT, Rajalakshmi Engineering College, Thandalam, Chennai, e-mail: babu.rajen17@gmail.com
Prithi Samuel, Department of CSE, Rajalakshmi Engineering College, Thandalam, Chennai, e-mail: drprithisamuel@gmail.com
K. Jayashree, Department of CSE, Rajalakshmi Engineering College, Thandalam, Chennai, e-mail: k.jayashri@gmail.com

https://doi.org/10.1515/9783110750584-009

in healthcare [1]. Cognitive computing is a subset of artificial intelligence that performs massively parallel processing to draw better decisions than humans and enables the interaction with fellow devices to deduce contextual interpretation based on past experience and proposes possible inferences and suitable solutions [2]. Progress in cognitive technologies has resulted in better performance, improved customer service, patient satisfaction, patient management, and automatic prior authorization process. The cognitive computing in healthcare accepts different types of structured and unstructured data as input and produces possible probabilistic inferences as output as shown in Figure 1.

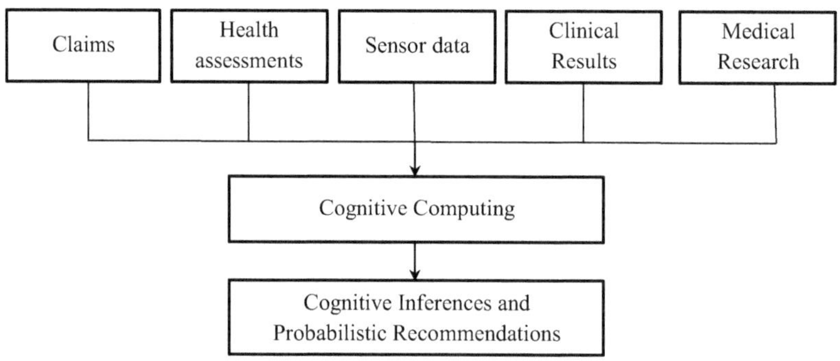

Figure 1: Cognitive computing in healthcare.

Cognitive computing is the key enabler for monitoring health records from sensors virtually using AI and robotics for efficient treatment and diagnosis when compared with an intensive human labor under critical illness such as chronic and psychological diseases [21]. The Health Insurance Portability and Accountability Act of 1996 mandated the necessity to identify and manage authentication policies to access sensitive health records of participants [3]. Meena and Gayathri [4] proposed a multi-factor authentication mechanism to guarantee the safety, confidentiality, and data protection of important patient information stored in the cloud. Hybrid methods for encryption, decryption, and encryption can be used to strengthen and safeguard the security of health records. Several security features such as integrity, authentication, access control, decryption, encryption, privacy protection, and key management must be considered in cognitive smart healthcare services [5]. Cognitive computing facilitates patient-oriented cognitive intelligent distribution of various computing resources depending on the current risk condition of each patient [6]. The cognitive computing system involves recent major technologies such as natural language processing to handle multistructured data, big data storage, and analytics to hold a huge variety of data, and Machine Learning (ML) algorithms to classify data in different forms.

The authentication factor plays a major part in safeguarding the healthcare system and it is critical to establish secure communication among sensor-enabled handheld devices and smart cards utilized by different participants [7]. Biometric-based authentication is also used to secure critical healthcare data. The adversary might affect the authorization, authentication, and privacy of the cognitive healthcare system by denial of service, eavesdropping, replay, structured query language injection, and spoofing attacks [8, 22].

The organization of the chapter is as follows: Introduction provides an overview of cognitive computing. The next section emphasizes on cognitive authentication system in smart healthcare with descriptions about the system, network, and threat models. The following section provides description about the methods for cognitive authentication system for smart-based healthcare. This is followed by the section that discusses the benefits and limitations of cognitive authentication system for smart-based healthcare. Then resolution solutions for cognitive authentication framework is explained. Challenges and future research directions are discussed in the next section. The chapter concludes with the final section.

Cognitive Authentication System in Smart Healthcare

The increasing demand in understanding and applying the potential value of cognitive computing enhances the quality of health care urge researchers to identify and formulate the authentication schemes necessary for secured and private access of sensor data [9, 23]. The cognitive authentication framework incorporates essential principles of privacy and security process data from various sensors and performs deep learning (DL) and ML-based classification to monitor and maintain different forms of authentication, and thus guaranteeing that the reliable healthcare service is offered to the patients and healthcare professionals [10]. The healthcare authentication system for cognitive computing encompasses system, threat, and network models.

System Model

The network-enabled devices in wireless computing for the healthcare domain can communicate without any interruption through Wireless Medical Sensor Network (WMSN) with low battery power and limited latency. With the power of IoT-enabled healthcare service delivery, the sensors embedded in the patient body can gather data like visual information, temperature, oxygen saturation level, pressure, imaging from various sensors, and communicate it to the gateway node as shown in Figure 2.

The system model consists of four participants namely healthcare professionals like physician, nurse and doctor, gateway node for accessing, certificate authority for ensuring healthcare service delivery, and collection of sensors inside the patient's body. The gateway node is responsible for broadcasting location information, patient data attained from sensors upon validation of the illegitimate sensors and patient credentials. The center for registration certificate authority is highly accountable for any trust issues related to privacy and security [11, 25].

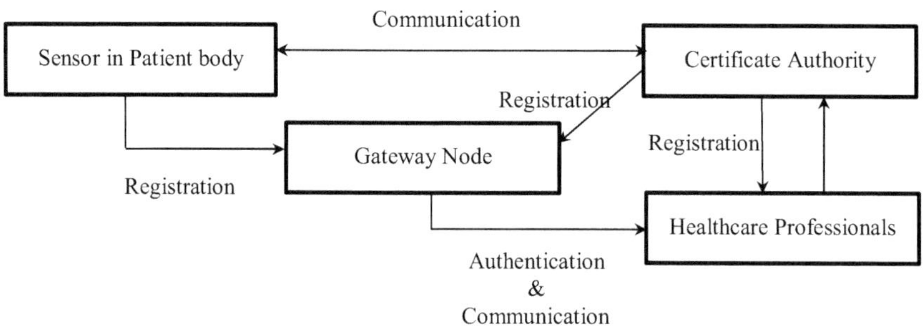

Figure 2: System model.

Threat Model

This model figures out the possibilities of an attack that might happen between two participant's communications via an open network channel. The adversary might obtain the critical credentials, extract the sensitive data from the storage, drop, alter, update, damage, or insert false data by intruding the patients or physicians sensor or mobile devices. In cognitive computing, threat model encompasses the following responsibilities.

Privacy Threat

The adversary uses some software to extract location and other personal information and utilizes some other software for decrypting and de-authenticating the message thereby disrupting the normal operation by flooding the communication channel.

Spoofing Threat Identification

The attacker might intrude the communication network to obtain the credential of a legitimate user in the system and might control the communication channel by altering, deleting, or injecting false data during transmission.

Traffic Analysis Threat

The adversary might take control over the broadcasting data from sensors, identifies the traffic if any, and utilizes that situation to tamper the data. Depending on the nature of tampered data, the attacker might use it for threatening the participant.

Access Control Threat

In cognitive computing environment the adversary can gain knowledge on various policies and can gain complete control over the communication channel by analyzing overall system activities.

Stolen Verifier Threat

The adversary can get the details of a healthcare professional or patient's sensor from the participants stolen mobile or sensor memory. The attacker can then decrypt the data and begin accessing data using the authentication credentials obtained.

Network Model

In network model, the nodes are the sensors, gateway node, certificate authority, and healthcare professionals and the links are the communication channels between different participants. The general procedure is detailed further.

All the participants need to register in the certificate authority for proper authenticated communication.

Different sensors placed inside the patient body will continuously monitor, gather data, and broadcast the same to the gateway node via WMSN.

From WMSN in the gateway node, the information is transmitted to the target healthcare professional for possible diagnosis.

Methods for Cognitive Authentication System for Smart-Based Healthcare

The healthcare enterprise is nowadays a rapidly developing field by means of extensive needs. It does not simply deliver vital as well as perilous facilities to the sufferers; it furthermore leads to huge incomes toward the health division. Healthcare workers strive in delivering reliable as well as economical medicare facilities to smart town people [12]. Various classy and economical instruments have been developed for providing services such as checking cortisol, insulin, blood pressure, body temperature, tension level, electrocardiogram (ECG), and electroencephalogram (EEG). Real-time indicators, namely EEG, that are attained by means of canny sensors as well as IoT instruments are typically multifarious, hence it needs progressive methods, namely ML, deep learning (DL), and big data analytics (BDA). All methods such as AL, ML, and BDA afford the essential handling as well as storing ability aimed at data storing, handling, and exploration. Thus the various methods used for cognitive authentication system for smart-based healthcare are shown in Figure 3.

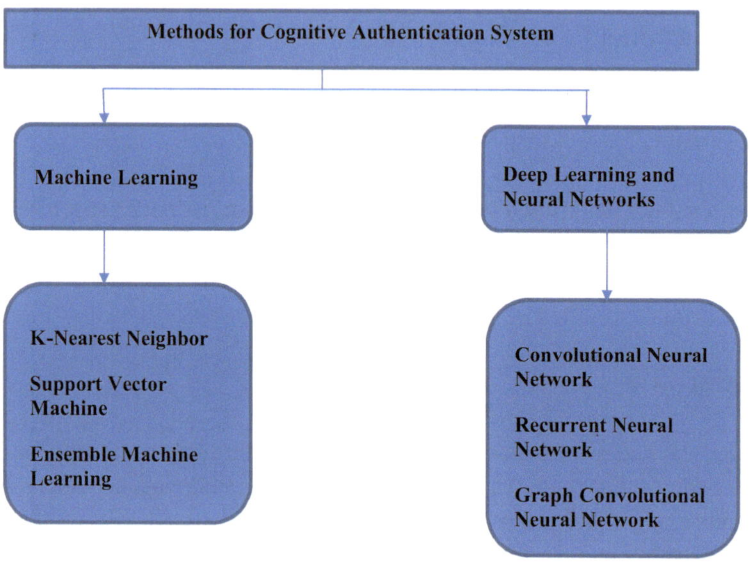

Figure 3: Methods for cognitive authentication system for smart-based healthcare.

Machine Learning

The various ML methods used for smart-based healthcare systems are described as given below.

K-Nearest Neighbors

KNN is a basic supervised learning technique such that the entities are categorized and centered on likeness toward specific characteristics of further entities of which group is fixed. Distances, typically Euclidean, are computed for those objects belonging to a group are bounded toward its KNN. KNN is useful for grouping of class data although the training dataset is tremendously trivial. It is also extensively used in several applications [26].

Support Vector Machine

SVM is a kind of supervised procedure which offers data exploration for classification as well as regression exploration. Initially, the SVM procedure was suggested for addressing twofold classification complications. Afterward the procedure has been used for addressing multiclass classification complications by means of diverse measures, by means of using individual SVM for distinguishing each class with all other classes through the one-versus-rest methodology.

SVM is built on the notion of taking a hyperplane which will well split features within dissimilar areas. In general, the purpose concerned using SVM is for reducing training flaws as well as to support the simplification facility toward different hidden data grounded on the standard of physical risk step-down. SVM is proposed for finding a hypothesis (I) as of the hypothesis space (H) which will ensure the least likelihood of flaw with the specified training sample in connection with the normalized function. By the learning procedure, the user can get the flawless hyperplane which will split the data samples. In the SVMs procedure the significant feature is using the kernel utilities for handling the nonlinearly independent data, which will transform the data toward higher dimensional space for grasping the nonlinear associations in the statistics parameters by linear hyperplanes that stays in the upper ranges [27].

The benefits of using SVM are such as assured optimal owing to overall least and not a limited least, easily implementable in Python, can be applied for linear as well as nonlinear severability such as smooth independent data cause rigid boundary, and the nonlinearly independent cause the low boundary, it provides the function plotting of elementary dot product by the use of kernel. The drawbacks are SVM that could not manipulate complete data hence it provides inferior solutions. Selecting the kernel is also the utmost drawback.

Ensemble Machine Learning

EML methods employ many learning techniques instead of a specific procedure to achieve better expectation accuracy. [24] have proposed a novel model that is centered on IoT and ML for monitoring the well-being situations of stroke concerned aged persons. They have suggested the prominence of observing patients distress as of prolonged infections as such, the doctors can give suitable medication. The suggested archetypal measures the present fitness pulse rate of sicked people with the use of specified sensors. The data captured from the sensor is transferred to the cloud for storage. When the parameters received are abnormal means, an urgency message as well as an email has been directed to the doctor or to the specific concierge. Likelihood archetypal has been made by various classifiers and gives the output as checking the patient requires any emergency needs. They have also discussed that their ensemble classifiers worked well when compared with the various classifiers.

Deep Learning and Neural Networks

DL architypes are used to decode EEG indications since the architypes is used to efficiently obtain the multiscale characteristic depiction of the indications plus the latent association of inner structures over and done with the nonlinear abysmal arrangement. EEG indications have compound statistics measureable plus solid time-variable in addition to uncertain structures, hence it is promising toward extracting an abysmal measurement as well as foremost feature depiction. Maiorana [13] utilized Siamese CNN toward acquiring abysmal structures for chore autonomous authentication. Furthermore, DL archetypal performs with the actual data openly and not having multifarious preprocessing as well as characteristic mining procedures.

Convolutional Neural Network

CNN is a deep neural network in association with a Gaussian kernel. As the Gaussian kernel is present, the convolutional layer has been partly associated with the preceding layer and receives variables that will significantly lessen the quantity of complex bounds. Various applications such as image processing use CNN and accomplished good results. Currently CNN has been utilized in analyzing as well as decoding EEG indications [14].

Recurrent Neural Network

Though CNN is capable of extracting three-dimensional features completely, EEG indications are extremely time-based as well as comprise rich timing attributes. The main features of EEG indications are the time-based features that imitate the variations in brain activities. For the time-based characteristics CNN will not provide accurate results, but RNN will provide tremendous time-dependent manifestation abilities. Hence, RNN has benefits in handling time-varying EEG indications [15].

Graph Convolutional Neural Network

GCNN outspreads the CNN toward the graph specialty. [18] have suggested an archetypal aimed at identifying the EEG biometric characteristics of diverse errands. The EEG indication record has been assessed centered with intra-frequency as well as cross-frequency efficient association. For the EEG indication record GCNN remained to spontaneously seizure the abysmal inner arrangement depiction. The GCNN also generates a symbolic illustration of EEG indications in addition to it spontaneously mine abysmal basic characteristics as of the lively EEG indications toward authentication. In the interim, the efficient combination among indications remained for solving the delinquent which will rely upon one variate features that aids GCNN for extracting inner basic characteristics.

Benefits and Limitations of Cognitive Authentication System for Smart-Based Healthcare

EEG-based perceptive classifications including genetic structures remain important in choosing an appropriate procedure for monitoring as well as identifying the uniqueness of people for the enhanced as well as robust security intensities.

Benefits

Cognitive methodologies encompass more benefits when compared with the conventional methods of biometrics.

Objectivity: With the significance of independence it is promising to create calculations which will not hinge on the consumer's view. Independence remains significant features which will surge the consistency of a dimension method.

Multidimensionality: Multivariate methods remain capable in the direction of providing diverse faces of customer circumstances.

Restrained: Though perceptive methods need the cathodes to be placed on the body, they will not openly obstruct using consumer errands corresponding auxiliary chore methods. Hence, they remain deliberated as restrained methods.
Obliqueness: By means of associating the main chore methods, the aforesaid methods do not need the dimension of manifest portrayal. They deliver the required data discreetly.
Stability as well as receptiveness: Psychophysical actions occur as uninterrupted movements hence it is used in real time. Therefore, discerning deviations once they arise in retort toward managements based on customer inducements remains potential.

Limitations

Distinct instruments: Cerebral motions remain computed with distinct instruments hence it will be costly. Further, diverse cathodes to be placed on the body remain applied in the particular dimensions. The assortment as well as precise arrangement of the cathodes remains critical to obtain data without noise. Hence, adequate considerations as well as time have to be provided to worker's preparation and equipment sustentation.
Information acquirement and analysis: Psychophysiological actions remain typically feeble electrical indications hence it is very much predisposed toward noise. Hence, appropriate pass through some filter methods has to be designated to apply for data. The features such as ambient lightning, ERP as well as ECG have to be considered. Data analysis remains an additional challenging for investigator as well as developers. This is due to the reasons which these methods generate huge volume of data which is challenging to examine.
Abnormality: Cathodes are linked to consumer through braid and the aforesaid detains the measure about a specific as well as disrupts the openness about the collaboration among the consumer plus the system.

Resolution Solutions for Cognitive Authentication Framework

EEG-based cognitive systems with biological aspects are important factors in determining the best procedure for monitoring and identifying people's identities for stronger and more resilient security. According to the discrete and distinct responsibilities for each biological attribute in the biometrics, noise and vibration develop between neurons when used for human recognition and detection. Furthermore, randomly gathered neurons are related to depression processes and superimposed behavior of other comparable brainwave patterns, which speeds up the processing time. Physical attributes like visual attention, pulse rate, and blood flow differ slightly from EEG signals. During capturing of EEG data, the massive and massive burden of brain reactions, distortion, and sound levels are rather low. As a result, certain rigorous and effective actions must be performed to eliminate the majority of the unnecessary sound in the system's highly sensitive data. The fundamental advantage of EEG-based biometric signals is that they have less noise and artifacts. Whenever limited hardware components are used with EEG-enabled biometrics, the level of security becomes less trustworthy. These security vulnerabilities are part of a broader bunch of problems in the healthcare sector.

Security Authentication Using Brainwaves

Authentication is useful in a variety of situations; however, its primary objective is to secure anything of value if users have access to a small number of people or organizations. The wide spectrum of authentication solutions now includes brain wave authentication mechanism. The electrical impulses of a human brain are employed to validate authenticity in this case. It has a number of benefits over credentials as well as other biometric such as fingerprints and retina scans. It is impervious to the "shoulder-surfing" problem that plagues password systems. Rather than entering a password, the user must memorize the unlocking associated with brain waves and create it in order to gain access. It also has a lower level of exposure than other biometrics such as fingerprints and retinas, ensuring greater integrity and validity.

With the rising popularity in therapeutic applications, the research establishment has discovered brain signals as a distinctive property that can be used in automated human biometric identification. Even with its resilience toward spoofing attacks, ability to undertake continual authentication, inherent real-time intrusion detection, and generality, neural impulses have several advantages over one of the most widely deployed biometrics, including facial, retina, and fingerprint. As a result, using brain signals for authentication is advantageous. There are also numerous issues that must be carefully addressed. The human computer interface system can collect equivalent electrical neural signals produced by interconnected neurons using brain wave detectors. Dry sensors and an auxiliary electrode comprise the associated brain sensor. The unprocessed data is transformed into packets when it is recorded. These packets of data are subsequently sent through Bluetooth technology communication to the threshold differentiator device.

In the grade differentiator unit, a particular sequence of brain waves is predefined as an unlocking sequence of neural signals for authorization. To gain access, the user must first create an authenticated pattern. The grade differentiator device receives brainwave activity in real-time basis and compares them to a predefined sequence to see if they match. Then it transmits the appropriate ASCII values toward the inspection module.

EEG-Based Biometrics

Scientists have recently been more interested in EEG-based biometric authentication. The EEG-based biometric technology can be used in important authenticating applications that demand more privacy and are hackable attacks. Authenticity and security systems in defense, military, investigations, traffic control, aeronautical, nuclear power plants, and other sectors where spoof-proof biometrics is critical. But compared with traditional biometric procedures, EEG data collection takes quicker and is more vulnerable to potential influence.

One of its primary distinguishing features is that EEG correlates to a concealed intellectual work formed by a stimulation that is not visible in any way [9]. When compared to standard biometric techniques, the comprehensive anonymity that is offered allow for the development to be significantly more reliable, and frequent spoofing attacks may not have been as feasible in this circumstance. Further benefit of EEG is that patterns are hard to replace - characterized by the fact underlying cognitive activities are unique to each person. That implies whenever a stimuli is delivered, every individual will interpret it in a unique way. These indicators can be influenced by factors such as emotion, psychological stress, and an individual's personal mental condition, making it far more difficult for hackers to get authorization from someone who is under duress [19]. This also indicates that cyberattacks aimed at mimicking and copying signals would be prohibited, and putting a person in danger to have authenticated will be inconceivable.

This can be addressed by employing techniques that shorten the length of time required to gather EEG data and limit the impact of the external disruptions on data collection. In hazardous contexts, customized techniques [20] for recording individual brain activation to audio/visual stimuli can be applied for efficient data gathering. Through the use of assistive devices for data gathering can drastically minimize EEG data collecting timeframe. Effective channel (electrode) feature selection techniques [17] can be used to include the fewest number of sensor nodes possible while still providing the most racially discriminatory general topic properties of people, reducing fixed period.

Challenges and Future Research Directions

Conventional biological factors do not meet the requirements of various disciplines, including healthcare institutions, commercial organizations, and academics. Biometric

monitoring, in relation to existing password authentication, is more practicable and distinctive because of its biological functions. The production of tissue or organs reduces the burden required to memorize additional credentials and protection codes. The following is the biometrics input monitoring from a human:
- Fingerprint scans: scanning and mapping of a person's biometrics.
- Handprint scan: Follows and captures the hand's full pattern.
- Voice recognition: Detecting and capturing real voice data and characteristics.

When discussing the concept of real-world applications, certain critical procedures should always be taken into account. To address the aforementioned specific needs, a technology of this type must fulfill several fundamental requirements.
1. *Ubiquitous*: To lessen the possibility of intelligence leaking, we require genetic components to assess the identification of information system.
2. *Individuality*: To produce an exact error-free outcome, each patient's physiological trait must be distinguished from others.
3. *Cohesion*: There would be far less volatility in the result and it must be relatively precise.
4. *Adaptability*: In terms of physiological properties, adaptability is flexible and extensible. As a kind of a consequence, extracting and malfunctioning the sensitive information will be difficult.
5. **Appropriateness:** Because of the sensitive nature of the data, it is critical to obtain underlying concept for monitoring and archiving.
6. **Reliability:** Organic qualities are much more difficult to copy, making them extremely durable in terms of addition to safety challenges.

The biometrics pattern(s) could be impacted in the case of emergency, wounds, or other incident difficulties, and people should not be able to undergo individual identity verification until their ailments have recovered. Fingerprints could still be conducted in a state of drowsiness and quiescent state, which is a risk and restriction. It would still be difficult to be commandeered and disclosed if fingerprints are cross-checked with the EEG signal. In contrast to biometric identification, EEG-based systems face a significant obstacle in the context of Gaussian noise that might compromise important healthcare data. As a result, researchers should be cognizant of both the benefits and drawbacks of integrated thermocouples.

Conclusion

This chapter outlined present condition approaches and suggestions for addressing privacy and security issues in smart healthcare system by presenting an EEG-driven reliable and secure cognitive authentication system. By means of the financial wealth plus technical development, mobile healthcare uses remain increasing. Smart instruments are in extensive usage for human fitness checking. The capable exploration plus analysis of fitness data faces several problems. Both researchers and professionals investigate and confront the most notable and significant EEG supported biometric solutions from a practical perspective. The broader scope and impact of the EEG-driven biometric system has been investigated and predicted to be deployed as a recommendation and paradigms setting technique for next-generation systems. Future biometric environments must be built in such a way that EEG television stations, technology, and process control techniques are all intimately integrated and synchronized. With today's global healthcare market in mind, it has been discovered that energy optimization and nodes integrity throughout sharing of information are key elements.

References

[1] Z. Ashfaq, A. Rafay, R. Mumtaz, S. M. H. Zaidi, H. Saleem, S. A. R. Zaidi, S. Mumtaz, and A. Haque, "A review of enabling technologies for Internet of Medical Things (IoMT) Ecosystem," Ain Shams Engineering Journal, vol. 13, no. 4, pp. 101660, June 2022. doi: https://doi.org/10.1016/j.asej.2021.101660.

[2] S. Gupta, A. Kar, A. Baabdullah, and W. Al-Khowaiter, "Big data with cognitive computing: A review for the future," International Journal of Information Management, vol. 42, pp. 78–89, 2018. doi: 10.1016/j.ijinfomgt.2018.06.005.

[3] A. Fruhling, D. Ramachandran, T. Bernard, R. Schuetzler, and J. Windle, "Patient preferences for authentication and security: A comparison study of younger and older patients," in Proceedings of the 2018 ACM SIGMIS Conference on Computers and People Research (SIGMIS-CPR'18). Association for Computing Machinery, New York, NY, USA, pp. 11–17, 2018. doi: https://doi.org/10.1145/3209626.3209702.

[4] S. Meena. and V. Gayathri, "Securing personal health records using advanced multi-factor authentication in cloud computing," International Journal of Recent Technology and Engineering, vol. 8, no. 6, pp. 5133–5140, 2020. doi: https://doi.org/10.35940/ijrte.f9724.038620.

[5] D. Rizk, H. Hosny, S. ElHorbety, and A. Salem, "SMART hospital management systems based on internet of things: Challenges, intelligent solutions and functional requirements," International Journal of Intelligent Computing and Information Sciences, vol. 22, no. 1, pp. 32–43, 2022. doi: 10.21608/ijicis.2021.82144.1107.

[6] G. D. Kumavat and S. Kumar, "Health monitoring using edge cognitive computing based smart health care," International Journal of Innovative Technology and Exploring Engineering, vol. 9, no. 1, 5356–5361, 2019.

[7] M. Jayabalan and T. O'Daniel, "A study on authentication factors in electronic health records," Journal of Applied Technology and Innovation (e-ISSN: 2600–7304), vol. 3, pp. 7–14, 2019.

[8] A. Ali, H. A. Rahim, M. F. Pasha, R. Dowsley, M. Masud, J. Ali, and M. Baz, "Security, privacy, and reliability in digital healthcare systems using blockchain," Electronics, vol. 10, no. 16, pp. 2034, 2021. doi: https://doi.org/10.3390/electronics10162034.

[9] A. Dhir, "The emerging role of cognitive computing in healthcare: A systematic literature review," International Journal of Medical Informatics, vol. 129, 2019. doi: 10.1016/j.ijmedinf.2019.04.024.
[10] A. H. Sodhro, C. Sennersten, and A. Ahmad, "Towards cognitive authentication for smart healthcare applications," Sensors, vol. 22, no. 6, pp. 2101, 2022. doi: https://doi.org/10.3390/s22062101.
[11] S. U. Jan, S. Ali, I. A. Abbasi, M. A. A. Mosleh, A. Alsanad, and H. Khattak, "Secure patient authentication framework in the healthcare system using wireless medical sensor networks," Journal of Healthcare Engineering, pp. 9954089, 22 July 2021. doi: 10.1155/2021/9954089. PMID: 34336174; PMCID: PMC8324348.
[12] G. Muhammad, S. K. M. M. Rahman, A. Alelaiwi, and A. Alamri, "Smart health solution integrating IoT and cloud: A case study of voice pathology monitoring," IEEE Communications Magazine, vol. 55, no. 1, pp. 69–73, January 2017 2017.
[13] E. Maiorana, "Learning deep features for task-independent EEG-based biometric verification," Pattern Recognition Letters, vol. 143, pp. 122–129, 2021.
[14] S. Zhang, L. Sun, X. Mao, C. Hu, and P. Liu, "Review Article Review on EEG-Based Authentication Technology Hindawi," Computational Intelligence and Neuroscience, vol. 2021, Article ID 5229576, pp. 20 pages, 2021. doi: https://doi.org/10.1155/2021/5229576.
[15] X. Zhang, L. Yao, C. Huang, T. Gu, Z. Yang, and Y. Liu, "DeepKey: A multimodal biometric authentication system via deep decoding gaits and brainwaves," ACM Transactions on Intelligent Systems and Technology, vol. 11, no. 4, pp. 49, 2020.
[16] A. J. Bidgoly, H. J. Bidgoly, and Z. Arezoumand, "A survey on methods and challenges in EEG based authentication," Computers & Security, vol. 93, pp. 101788, 2020.
[17] T. Alotaiby, F. E. A. El-Samie, S. A. Alshebeili et al., "A review of channel selection algorithms for EEG signal processing," EURASIP Journal on Advances in Signal Processing, vol. 2015, no. 1, pp. 66, 2015.
[18] M. Wang, H. El-Fiqi, J. Hu, and H. A. Abbass, "Convolutional neural networks using dynamic functional connectivity for EEG-based person identification in diverse human states," IEEE Transactions on Information Forensics and Security, vol. 14, no. 12, pp. 3259–3272, 2019.
[19] Q. Gui, Z. Jin, and W. Xu, "Exploring EEG-based biometrics for user identification and authentication," in Signal Processing in Medicine and Biology Symposium (SPMB), 2014.
[20] D. Rajesh Kumar, K. Rajkumar, K. Lalitha, and V. Dhanakoti (2020). "Bigdata in the management of diabetes mellitus treatment," in Studies in Big Data, Springer Singapore, pp. 293–324.
[21] S. Wan, Z. Gu, and Q. Ni, "Cognitive computing and wireless communications on the edge for healthcare service robots," Computer Communications, vol. 149, 2019. doi: 10.1016/j.comcom.2019.10.012.
[22] R. Sathish and D. R. Kumar, "Proficient algorithms for replication attack detection in Wireless Sensor Networks – A survey," in 2013 IEEE International Conference on Emerging Trends in Computing, Communication and Nanotechnology (ICECCN), March 2013.
[23] R. Sathish and D. R. Kumar, "Dynamic detection of clone attack in wireless sensor networks," in 2013 International Conference on Communication Systems and Network Technologies (CSNT 2013), April 2013.
[24] R. Ani, S. Krishna, N. Anju, M. S. Aslam, and O. S. Deepa, "IoT based patient monitoring and diagnostic prediction tool using ensemble classifier," in 2017 International Conference on Advances in Computing, Communications and Informatics (ICACCI), 2017, pp. 1588–1593.
[25] K. D. Rajesh and S. ManjupPriya, "Cloud based M-Healthcare emergency using SPOC," in 2013 Fifth International Conference on Advanced Computing (ICoAC), December 2013.

[26] H.M.K. Bharadwaj, A.Agarwal, V. Chamola, N.R Lakkaniga, V. Hassija, M.Guizani, and B. Sikdar, (2021) A Review on the Role of Machine Learning in Enabling IoT Based Healthcare Applications Special Section on AI and IOT Convergence for Smart Health IEEE Access Digital Object Identifier 10.1109/ACCESS.2021.3059858.
[27] C. Azad, A.K. Mehta, D. Mahto, D.K. Yadav, (2020) Support Vector Machine based eHealth Cloud System for Diabetes Classification, EAI Endorsed Transactions on Pervasive Health and Technology 01 2020 - 05 2020 | Volume 6 | Issue 22 | e3 doi: 10.4108/eai.13-7-2018.164627.

B. Lakshmanan, S. Priyadharsini, G. Priyanka

Cognitive-Inspired Computer Vision Assist System for Diabetic Retinopathy Detection from Fundus Images

Abstract: Diabetic retinopathy (DR) is one of the major reasons for blindness in humans. DR detection at the early stage is a very hard and challenging one. Rapid advancements in computer-aided systems necessitate an error-prone framework that can detect DR severity grading accurately. We attempt to design a cognitive-inspired intelligent tool for detecting DR from diabetic retinal fundus images. We presented novel framework which consists of stages of a deep light-weight convolution neural network (CNN) model that can be able to process and perform well even for low-quality diabetic retinopathy images. The proposed framework can be able to extract hard and soft exudates from the images very accurately for further processing using a small attention-based mechanism inspired by the small face attention network. The system is evaluated on the benchmark APTOS dataset and shows improved performance over traditional techniques.

Keywords: Cognitive-inspired deep learning model, convolutional neural network, diabetic retinopathy detection, severity grading system, small attention network

Introduction

Diabetic retinopathy (DR) is a serious illness caused by the blood vessels present in the retina of the human eye. Eye Diseases Prevalence Research Group study reveals that DR results in major eye diseases and disorders in humans, which causes blindness among the age group 20–75 years. A recent survey indicates that the DR infection rate among adults worldwide was estimated at around 103.12 million and gradually increase up to 160.50 million in the year 2045. The highest prevalence of DR was in Africa with 35.90% and it is followed by North America and the Caribbean which contributes around 33.30% and the lowest rate reported in South and Central America with 13.37%. The International Diabetes Federation estimated that recently that diabetes mellitus in humans also causes DR at several levels. Most people experience symptoms like the mutilation of blood vessels in the retina region of the human eye. The severity of DR leads to proliferative diabetic retinopathy which may cause fragile blood vessel growth in the retina. The advanced computer-

B. Lakshmanan, S. Priyadharsini, G. Priyanka, Department of Computer Science and Engineering, Mepco Schlenk Engineering College(Autonomous), Sivakasi 626005, India

https://doi.org/10.1515/9783110750584-010

assisted tool is designed to develop the severity grading system and lesion localization in fundus images. The cognitive-inspired machine vision system will assist ophthalmologists in diagnosis DR at the earlier stage. The framework will help to reduce the labor-intensive work and time-consuming process. Therefore, it necessitates the development of a cognitive-based computer-assisted tool based on a deep network model that can be used to extract discriminative descriptors from the given fundus images and also results in the severity of the DR grading for a preventive treatment plan. Figure 1 shows the different severity levels of DR fundus image.

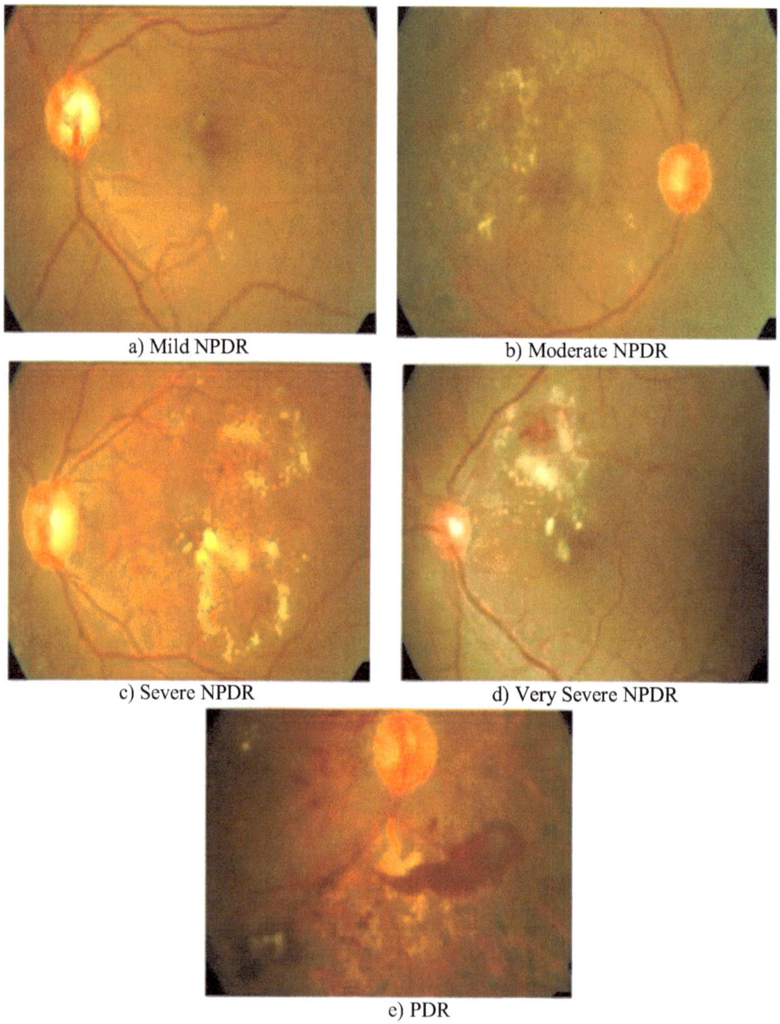

Figure 1: Different severity levels in DR images.

The key contributions of the proposed framework for diabetic retinopathy detection are:

1. Proposed a new custom model inspired by the small face attention network of MTCNN to detect fundus in the diabetic retinopathy images.
2. Compared the results with the benchmark dataset and the proposed system outperformed the existing computer-aided tools in terms of diabetic retinopathy detection in both high and low-quality retinal images.

The book chapter is structured as follows: The "Related works" section describes the related works of diabetic retinopathy. Materials and methods are explained in the next section, while the following section elaborates the experimental results and discussion, and the conclusion of the proposed work is presented in the final section.

Related Works

Most of the recent works related to fundus detection in diabetic retinopathy images are based on deep learning-based architectures. Khan et al. [1] proposed a VGG–NIN-based invariance to the scaling model by staking custom layers on top of VGG-16 to learn features effectively. To learn the data's nonlinear behavior (network-in-network), NiN was added at the top of the SPP layer. The softmax activation function was used for classification in the last layer. The categorical cross-entropy error function is used to measure the performance of the model. The stochastic-gradient descent optimizer was used to perform training. The proposed model was compared with the other state-of-the-art work in which the proposed model works better than the compared model. The obtained value for accuracy, precision, recall, specificity, and F1-score is 85%, 67%, 55%, 91%, and 59.6%. Gao et al. [2] proposed the automatic prediction of DR infection for providing valuable consultation to DR-affected patients. They have constructed their custom-labeled dataset for DR fundus images with the assistance of medical professionals. This constructed dataset is used for training the DL model. Their self-built dataset contains 4,407 images which are labeled as normal, moderate, heavy, and severe. The preprocessing step involves size, shape, and color normalization. After preprocessing and augmentation, the dataset is given as the input for the Inception-V3 deep neural network model. The experimental result was compared with the other works on similar tasks. This system achieves 88.72% accuracy, 96.51% precision, and 94.84% recall. The drawback of this work is that this proposed highly complex deep learning model needs a larger dataset for the training process with high-quality image data.

To overcome the complexity of processing sequential complex models addressed in [1, 2], Qummar et al. [3] introduced an ensemble DL model for DR detection. Here the publicly available Kaggle dataset of retina images was used to design ensembled DL models using Resnet50, InceptionV3, Xception, Dense121, and Dense169. As opposed to our proposed work, this system is tedious to work with the available CPU/ commodity GPU support.

The framework proposed in [4] used inherent fundus detection to categorize the disease by a screen-check mechanism. The steps involved in this framework predominantly focus on adjusting the luminance, lighting conditions, and contrast. Also to prevent misclassification blood vessels and optic disk were ignored during processing. For luminosity and contrast enhancement, gamma correction and CLAHE were used. For vessel extraction, principal component analysis and using convex hull transform. After applying background subtraction, lesions are identified based on the shape information extracted from the image, and class labels are fixed based on the number of lesions detected. For the evaluation, the fundus retinal images used were selected from four publicly available databases, namely STARE, DIARETDB1, DRIVE, retinopathy online challenge (ROCh), and MESSIDOR. But this algorithm was not adaptive for real-time scenarios as it involves a lot of human intervention.

Cynthia et al. [5] proposed a branching-based parallel processing deep learning framework to diagnose the five categories of DR-related issues. The dataset was created by collecting fundus and nonfundus images from the Shenzhen Si Bright Co. Ltd. (Shenzhen, Guangdong, China) with a total of 89,917 images for experimentation. Shallow CNN with different scaled input processing for DR image detection and prediction was designed in [6]. The model in [6] has N number of shallow CNNs to capture the image descriptors in different illumination conditions. Feature vectors produced from N models will be fed to the dense layer for further classification using softmax as an activation function. The diabetic retinopathy detection dataset was used to train the model which contains 35,126 labeled images. With $N = 5$, the accuracy of the model in the training phase is found to be around 80%. The accuracy of the proposed model increases to 92% when the dataset size increases up to 35,000.

Another multitask-based learner was introduced in [7]. This model contains one classification model and one regression model. The proposed multitask model was applied to two datasets: APTOS and EyePACS. In data preprocessing, a normalization scale of 0–1 is applied for all the pixels in the input image. The augmentation process was done to increase the dataset size with the support of basic affine transformation techniques. During the training phase, the classifier and regressor were independently fine-tuned to accomplish the multiclass classification problem using the APTOS and EyePACS datasets. Additionally, the receiver operating characteristic curve for minimum and maximum area's average coverage was found to be 0.96 and 0.93, respectively. But as compared to the existing literature from [5–7] the proposed deep learning-based system can work with poor/low-quality retinal images.

Another deep learning-based diabetic retinopathy severity prediction model was introduced in [8]. In this model, feature extraction is performed from DR images and the extracted features are aggregated. The pretrained VGG-16 model is used for feature extraction. VGG-16 is composed of many convolution and pooling layers which help for feature extraction. Fusion methodologies are applied for performing feature aggregation. Feature aggregation results in a compact representation of extracted DR image features. These features are discriminative and are devoid of redundant information and noise. The severity level of DR is identified by training the aggregated features through DNN. Kaggle APTOS 2019 dataset was used in this paper for experimental analysis. This method stated that they have achieved 84.31% of accuracy. The limitation of this work is that the model is fully based on the existing complex layer configuration of VGG-16.

DR-grading classification using DCNN was proposed in [9] which classifies based on the hierarchical levels in DR grading. The generic and specific levels in the DR grading are preserved by employing two different classification networks. The coarse network predicts whether the given input retinal image is infected or not. Whereas the fine network acts as a multiclass problem to predict the continuous possibility of infection with different severity levels ranging from 1 to 5. In this work, it is developed as an interdependent counterpart. The authors performed experimentation on their datasets and inferred that their proposed CF-DRNet is classifying DR severity levels with superior performance when compared to the earlier works. But this system is based on two different handshaking networks to accomplish its task.

DcardNet: tomography-based diabetic retinal image processing at multiple levels was proposed in [10]. This work employed CNN to learn the relationship between the structural and angiographic optimal coherence. To avoid overfitting in the relationship learning process, the network was designed with an adaptive dropout rate and adaptive label smoothing mechanism. This work followed three severity grading levels for classifying the fundus images based on the International Clinical Diabetic Retinopathy scale. There are three stages of classification. In stage 1, the model predicts the presence of referral/nonreferral DR. The second stage classifies referral DR, if any, as nonretinopathy, nonproliferative retinopathy, or proliferative retinopathy. In the final stage, a fine classification is performed based on a severity scale such as mild, moderate, and severe for both proliferative and nonproliferative retinopathy stages. The authors applied k-fold cross-validation with 90% of the data for training and 10% of the data for assessing the trained model's accuracy.

Hybrid approaches are adapted to predict DR automatically from retinal fundus images and to classify the different stages of DR in [11]. In this work, the authors separated the landmarks of the retina and lesions from the input fundus images. Different segmentation algorithms are applied as preprocessing steps to detect and segment the portion of the retina from the retinopathy image. Features are extracted from the segmented portions and are fed to machine learning algorithms for performing DR classification. The classification results are validated, and further training fine-tunes the results. Experimental analysis is carried out and their achieved accuracy is 88.63%, SE is 86.25%, and the SPE is 91%.

Active deep learning-based diabetic retinal image processing was proposed in [12]. In this work, a multilayer CNN architecture is employed for classifying the fundus images and to predict diabetic retinopathy. The CNN model used in this work is an adaptive deep learning system that automatically extracts dominant features through active layers in the model. The ground truth labels associated with training images assist CNN to extract discriminative regions by learning both simple and intricate features. The severity levels of retinopathy are graded after annotation and the region segmentation of eye sample images.

A fine-grained annotated DR dataset was developed for retinal image disease detection [13]. This dataset is huge and consists of 2,842 images. Out of 2,842 images, 1,842 images have pixel-based lesion annotations, and the remaining 1,000 images have image-based DR annotations. The labels are graded by experts in the field of medicine. This large dataset will be needful for extensive research on diabetic retinopathy analysis. For evaluation purposes, three benchmark tasks are established such as (1) segmentation of lesions from DR images; (2) classification and segmentation of joints for grading DR images; (3) multi-disease identification by pretrained networks.

Most of the aforementioned literatures were based on the very complex deep learning model for automatic prediction of DR and classification of severity levels. But Gayathri et al. [14] developed a novel feature learning system using the CNN model to provide improved system performance. They have developed a simple CNN model for both binary and multiclass classification. CNN learns the important features from the input retina images and classifies the DR grading effectively compared to the machine learning models which work on hand-crafted features. Haralick and anisotropic dual-tree complex wavelet transform features were extracted for developing an effective and most reliable DR classification model in [15]. But our proposed system has a more reliable classification of DR with fast model convergence using an early stopping mechanism to avoid memorization issues. Also, with the invention of a custom CNN model with learnable parameters, the retinal image is efficiently processed in the proposed system as discussed in detail in the subsequent sections.

Materials and Methods

Dataset Description

In our work, APTOS 2019 dataset is used for severity grading identification and lesion detection. The dataset contains 3,662 training images and 1,928 test images with different dimensions. In this, the severity of diabetic retinopathy ranges from

0 to 4 in which “No severity” is assigned to value “0,” “Mild severity” is labeled with a value “1,” whereas “Moderate” and “Severe” are assigned with values “2” and “3,” and “proliferative DR” is labeled with a value “4.”

Preprocessing

Before training the images using convolutional neural network model, the images are preprocessed by reshaping it into uniform dimension (340 × 340). After reshaping, color enhancement is performed using histogram specification technique. The preprocessed images are then fed as an input to the deep CNN model. Here training images are spilt as k-fold (5-fold) for 50 epochs.

Convolution-Based Small Attention Network

The system design in Figure 2 gives an idea about the scheme by which the proposed system can be implemented. This will provide the overall flow of work and explains the various modules. The small attention network is designed by using a base network called multitask cascaded convolutional neural networks (MTCNN). MTCNN is a neural network originally designed for detecting small faces in input face images. MTCNN is popular because it has achieved high-quality results in bench-top sets, and it is able to process facial features in small regions. In the proposed work, we have adopted MTCNN because of detecting and localizing small, affected regions in the retinopathy images.

In the proposed modified network consists of two convolutional layers followed by two max pooling layers, two dense layers with Relu activation function are added at the last of MTCNN and the last layer contains one dense layer with softmax function. Finally, the model performance is evaluated by using the parameters, model accuracy, and model loss. Modified MTCNN-based small attention algorithm consists of three phases:

- In Phase 1, an image pyramid is constructed by reducing the scale of images iteratively. This step is important to extract features from very small regions.
- In Phase 2, the convolution neural network is trained using the dataset images. In addition, the scaled down images are augmented with the existing dataset to improve the results of smaller regions.
- In Phase 3, the detected features are clustered to localize the affected region.

In the proposed small attention network architecture is shown in Figure 3. The network learns multi-branch features in the initial layers that focus on low-level features. Next, maps of neighboring branches are inserted so that features from a larger scale can detect complex features on a smaller scale. Finally, a multi-level training

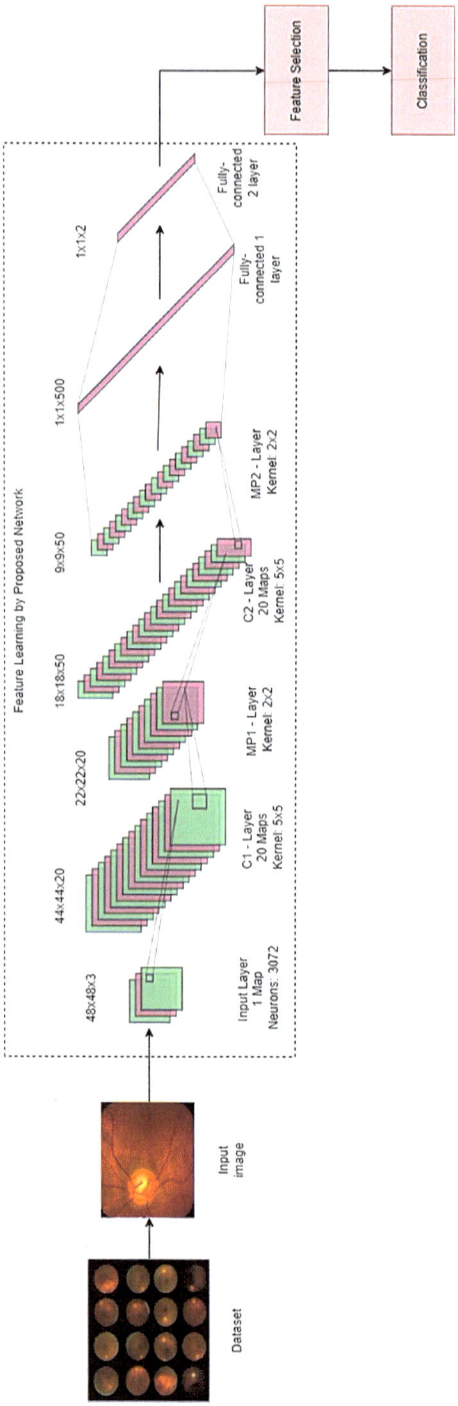

Figure 2: Proposed DR detection framework.

and testing are performed to make our model stronger by moving to a different level. Complete training is given to the network using different scaled images. After training phase, small attention network can detect and localize affected regions in the input image, especially on small retina portions of the image.

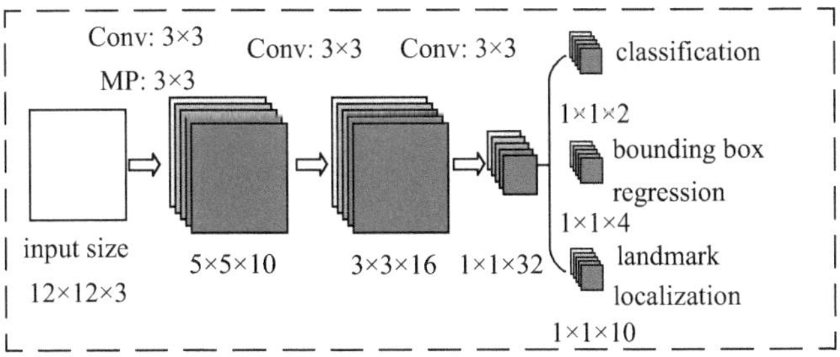

Figure 3: Small attention network architecture.

Experimental Setup and Results Discussion

The proposed DR framework has been developed using Python Keras environment using Tensor flow in the backend. The retinopathy in the APTOS 2019 dataset is given as input after preprocessing and data augmentation. The dataset images are high dimensional of varying size ranging from 1,050 to 3,216 pixels in both dimensions in PNG format. The images are color enhanced by improving the color values in red, green, and blue channels. Since color is the basic feature, color enhancement improved the network accuracy. Without GPU support, the high-dimensional images cannot be processed as it is. Hence the original images are then reshaped 340 × 340 dimension. In order to learn smaller portions, the reshaped image is scale-downed into different scales up to 16 × 16 pixels. This would result in increased volume of sample images for experimental purpose.

The proposed small attention network model made up of an input layer followed by five blocks of network. Each block consists of two to four convolution layers and one max pooling layer. The convolution operation in the convolution layer replaces conventional matrix multiplication, thereby learning important features from the high dimensional images in an efficient manner.

The weight updating happens in each block of the network. The network adapts itself by continuously updating weights to learn the input and output mapping relationship. The weight updating in the first block of the network is visualized in Figure 4. Figure 4 shows a 3 × 3 filter matrix used for convolution operation in the convolution layer of block 1. The white regions denote that the corresponding pixel's weights are being updated, gray regions denote that the corresponding weights are not updated, and black regions denote that the corresponding pixel weights are ignored. Thus, during

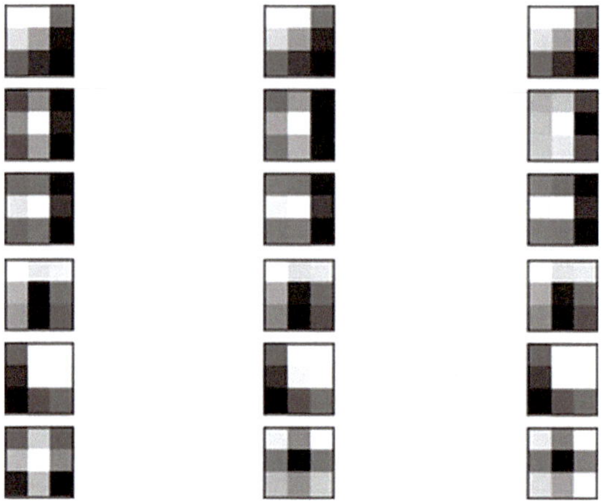

Figure 4: Weight/filter update visualization for Block1_Conv1.

convolution operation, the pixels convoluted with white pixels will have more importance/significance as compared to gray and black-colored weight values as illustrated.

The features learnt in the first convolution block is shown in Figure 5. Figures 6 and 7 show the feature learning in small regions of an image. Figure 6 shows feature learning in the first 200 × 200 image region, whereas Figure 7 shows feature learning

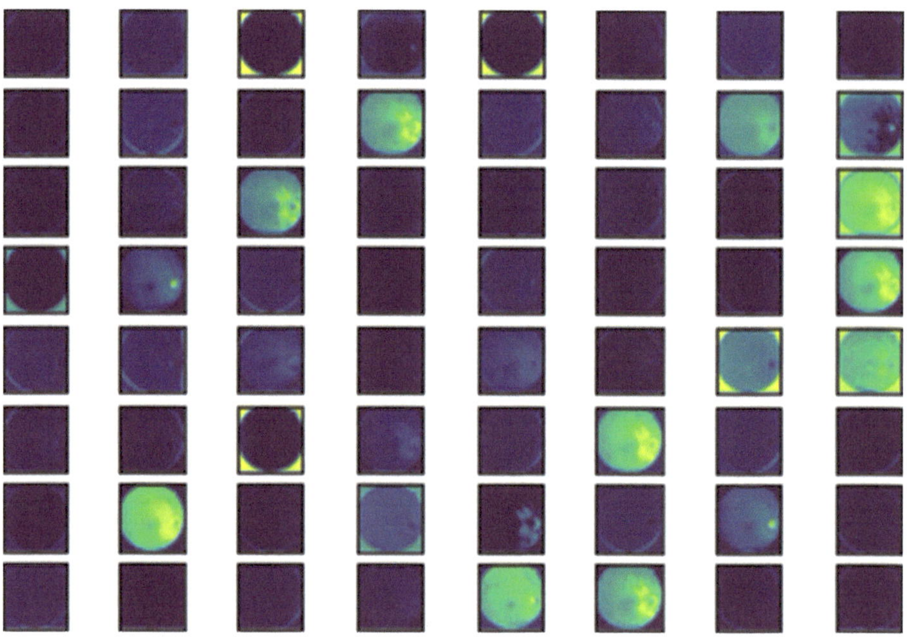

Figure 5: Feature visualization for the layer: block1_conv2.

Table 1: Training, validation, and testing accuracy for proposed network.

Accuracy	Train–validation–test split		
	60:20:20	**70:10:20**	**80:10:10**
Training accuracy	74.88%	89.12%	91.84%
Validation accuracy	86.39%	91.63%	92.37%
Testing accuracy	87.45%	94.81%	96.36%

in the first 32 × 32 image region. Figures 6 and 7 prove that the proposed small attention network learns features from small regions in an image.

The proposed network model is trained for 25 epochs after which the network is stabilized. The trained network is then tested on test data for the severity level

Figure 6: Feature learning in iterations illustrated in 200*200 portion.

identification of diabetic retinopathy, and it is observed that the network achieves the testing accuracy of 96.36%. The training, validation, and testing accuracy are tabulated in Table 1 for different ratio split.

Conclusion and Future Work

The proposed DR framework was used to detect the lesions in retina images. The work is tested on the APTOS 2019 dataset of high-dimensional retinopathy images. It is observed that novel small attention network will be able to learn the small features in the image which facilitates to detect small lesion region in the diabetic retinopathy image. The proposed network uses MTCNN as the base network. The modified layers

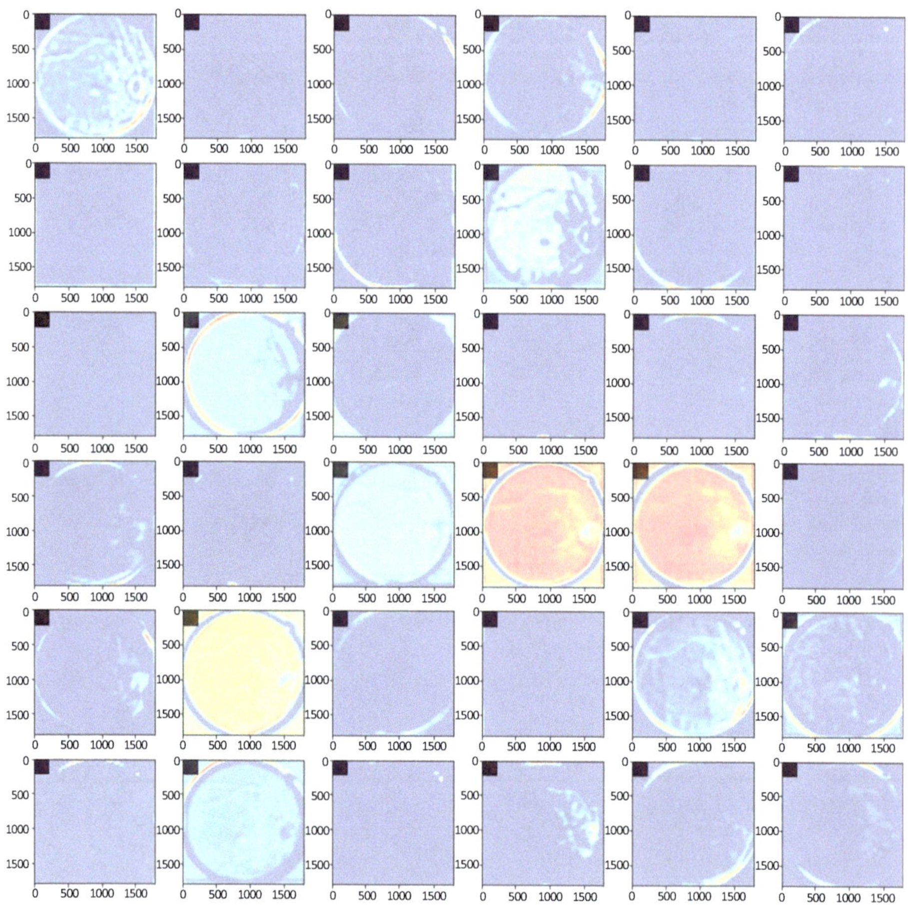

Figure 7: Feature learning in iterations illustrated in 25*25 portion.

in the network and the scaled down images constructed in the image pyramid construction made the network to learn features from very small regions of the image. Finally, the proposed network can report severity level of DR from the given input fundus images. The experimental results of the proposed framework were compared with existing pretrained models, and it is noted that the proposed DR framework outperforms the pretrained models.

In future, metaheuristics and bioinspired optimization algorithm-based feature selection can be incorporated to provide better solution with improved performance. After predicting the severity of the DR (0, 1, 2, 3, and 4), the number of lesions may be determined by using the count number of regular cells and odd cells. Model ensembles can be tried for classifying the DR to get higher consequences.

References

[1] Z. Khan, F. G. Khan, A. Khan, Z. U. Rehman, S. Shah, S. Qummar, F. Ali, and S. Pack, “Diabetic retinopathy detection using VGG-NIN a deep learning architecture,” IEEE Access, vol. 9, pp. 61408–61416. 2021.

[2] Z. Gao, J. Li, J. Guo, Y. Chen, Z. Yi, and J. Zhong, “Diagnosis of diabetic retinopathy using deep neural networks,” IEEE Access, vol. 4, pp. 3360–3370, 2019.

[3] S. Qummar, F. G. Khan, S. Shah, A. Khan, S. Shamshirband, Z. U. Rehman, I. A. Khan, and W. Jadoon, “A deep learning ensemble approach for diabetic retinopathy detection,” IEEE Access, vol. 7, pp. 150530–150539. 2019.

[4] J. Kanimozhi, P. Vasuki, and S. Md. Roomi, “Fundus image lesion detection algorithm for diabetic retinopathy screening,” Journal of Ambient Intelligence and Humanized Computing, Springer, vol. 12, pp. 7407–7416, 2021.

[5] J. Cynthia, M. Sankari, M. Suguna, and D. R. Kumar, “Survey on disaster management using VANET,” in 2018 4th International Conference on Computing Communication and Automation (ICCCA), December 2018.

[6] W. Chen, B. Yang, J. Li, and J. Wang, “An approach to detecting diabetic retinopathy based on integrated shallow convolutional neural networks,” IEEE Access, vol. 8, pp. 178552–178562, 2020.

[7] K. Sathya and D. R. Kumar, “Energy efficient clustering in sensor networks using Cluster Manager,” in 2012 International Conference on Computing, Communication and Applications (ICCCA), February 2012.

[8] J.D.Bodapati, N.S.Shaik, and V.Naralasetti, “Deep convolution feature aggregation: an application to diabetic retinopathy severity level prediction”, Signal, Image and Video Processing, vol.15, pp.923–930, 2021.

[9] W. Zhan, G. Shi, Y. Chen, F. Shi, X. Chen, G. Coatrieux, J. Yang, L. Luo, and S. Li, “Coarse-to-fine classification for diabetic retinopathy grading using convolutional neural network”, in Artificial Intelligence in Medicine. Elsevier, vol. 108, pp. 1–9, 2020.

[10] W. Chen, B. Yang, J. Li, and J. Wang, “DcardNet: Diabetic retinopathy classification at multiple levels based on structural and angiographic optical coherence tomography,” IEEE Transactions on Biomedical Engineering, vol. 68, no. 6, pp. 1859–1870, 2021.

[11] S. Karkuzhali and D. Manimegalai, "Distinguishing proof of diabetic retinopathy detection by hybrid approaches in two-dimensional retinal fundus images," Journal of Medical Systems, Springer, vol. 43, pp. 1–12, 2019.
[12] I. Qureshi, J. Ma, and Q. Abbas, "Diabetic retinopathy detection and stage classification in eye fundus images using active deep learning," Multimedia Tools and Applications, Springer, vol. 80, pp. 11691–11721, 2021.
[13] Y. Zhou, B. Wang, L. Huang, S. Cui, and L. Shao, "A benchmark for studying diabetic retinopathy: Segmentation, grading, and transferability," IEEE Transactions on Medical Imaging, vol. 40, no. 3, pp. 818–828, 2021.
[14] S. Gayathri, V. P. Gopi, and P. Palanisamy, "A lightweight CNN for Diabetic Retinopathy classification from fundus images," Biomedical Signal Processing and Control, vol. 62, pp. 1–11, 2020.
[15] S. Gayathri, A. K. Krishna, V. P. Gopi, and P. Palanisamy, "Automated binary and multiclass classification of diabetic retinopathy using haralick and multiresolution features," IEEE Access, vol. 8, pp. 57497–57504, 2020.

Dineshnath G, S. Saraswathi

A Novel Deep Belief Neural Network Model for Abstractive Text Summarization

Abstract: Everyday researchers present their research works in various formats such as research articles, reports, patents, and study chapters. All over the world, the number of submissions increases due to technology development that leads to innovations. Search results will provide numerous relevant results for a single word and it is quite difficult for beginners and also for experts to analyze the research work by reading the entire article. Text summarization has evolved to reduce this issue. It automatically examines the articles and extracts the important information from the research works. Due to specific technical aspects, conventional text summarization methods cannot be employed in the text summarization of research works. Researchers pay more attention to bring efficient scientific article text summarization and this research work aims to achieve the goal of text summarization using a deep belief neural network. Abstractive text summarization is proposed in the research work by addressing the importance of the research article summarization. This novel approach attains better performance than conventional text summarization models. The experiental results shows the superior performance of our approach.

Keywords: Abstractive text summarization, deep learning network, bidirectional encoder representations from transformers (BERT), natural language processing (NLP)

Introduction

The tremendous development of the web reduces the burden to the end user in information collection by providing relevant results for their queries. Further aid from state-of-the-art communication technologies has accelerated the rate of information acquisition. At high rates, there always exists a threat of dealing with a huge volume of data in which heterogeneity is more observed owing to the varying source of data as well as their acquisition mechanisms. In recent times, the provision of data from a large pool of data is a challenging task and manual search-based methods existing conventionally are quite tedious and time-consuming. Hence, it is essential to present relevant essential information to the user through automatic techniques

Dineshnath G, Department of Computer Science Engineering, PEC, Pondichery
S. Saraswathi, Department of Information Technology, PEC, Pondichery

https://doi.org/10.1515/9783110750584-011

which are considered as most important to meet out the user quality of service (QoS) criteria. Generally, for a single query of a user, multiple relevant results are presented which takes more time to read and select the essential information. Moreover, this process is tedious which makes the reader to quit the process. To overcome this issue, text summarization has been introduced. Basically, it is an NLP technique that provides a summary of texts from a sole document or multiple documents which considers only important features. A summarization model needs to bring all the salient information. The highlights of text summarization are (1) retrieval of essential information from long text, (2) quick processing, (3) easy to use, (4) rapid update of information, and (5) high-resolution results. Later, automatic text summarization is introduced to improve the above features by processing the documents using a computer program to obtain the important points in a summarized manner. Automatic text summarization helps to obtain a meaningful interpretation or a gist of information from a variety of data sources. This aids in improving the QoS especially in areas of Big Data and Cloud computing which are essentially concerned with the effective provision of services required by users.

Automatic text summarization systems provide a short description summary of documents considering the salient information for the user query [1]. If the query has multiple results, automatic text summarization clusters similar documents and presents the results in a clustered format. Figure 1 depicts the categorization of text summarization in detail.

The summarization of texts is widely utilized in search engines and news websites [2] Generally, text summarization is categorized based on the documents, summary context, function, type of summarizer, genre, and context type [3, 4]. Table 1 provides a comparative analysis of text summarization techniques as a summary.

Among the above text summarization techniques, generated summary-based text summarization gains more attention toward researchers. Since other than summary-based approaches all the summarization techniques depend on particular factors like input documents, genre, query-specific, and domain-specific requirements. But summary-based models are output-specific models that provide summarized results for the given data. Further, summary-based models are categorized into extractive and abstractive summarization [5]. In the extractive summarization, some portions from the original text are extracted based on the computed scores. The scores can be obtained considering the linguistic features or statistical features. In the case of abstractive summarization, the original text is rephrased and generated as a new phrase. The meaning of the generated phrase must reflect the content of the actual text. Abstractive summarization is quite difficult as it requires deep knowledge about the document to generate the summary [6]. Extensive NLP techniques and advanced machine learning methodologies are required for generation and or computation of abstractive text summarization. Though abstractive summarization is a complex process, it is far better than extractive summarization as the summary is quite alike a summary generated by a human [7]. This research work

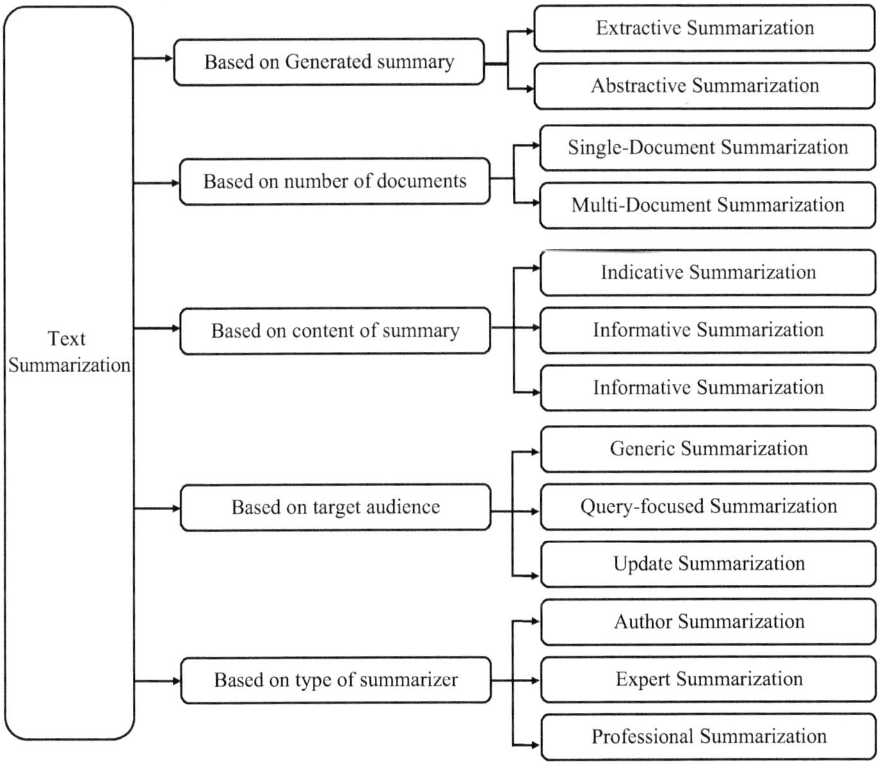

Figure 1: Categorization of text summarization.

Table 1: Text summarization techniques comparative summary.

Type	Description	Merits	Demerits
Abstractive	Text summarization process that creates a semantically related summary	Better compression ratio, semantically related summary	Computational complexities
Extractive	Text summarization process that creates a summary based on statistical features	Easy to compute	Inconsistency, imbalance nature, lengthy summary
Indicative	Present the main idea of the user	Quick categorization and easy to process	Less detailed information
Informative	Provides concise information about the given text	It can be used as a substitute for the main document	Could not able to provide a quick review

Table 1 (continued)

Type	Description	Merits	Demerits
Generic	Generalized summary irrespective of the user type	Can be used at any type of user applications	Provides authors view not user specific view
Query	Extracts information based on the user query	Specific information can be searched	Depends on user requirements, not a common summarization process
Domain	Depends on the user specified domain the summarization is processed	Aware of domain and dependent	Limited to subject of the documents
Genere	Accepts special type of text as input	Summarize heterogeneous documents	Text templates reduce the performance

focused on the abstractive text summarization approach. Before that, the limitations identified in the summarization process are presented as follows.

- Content selection: A document has vast information in it and all the information are not important as few texts are redundant [8], less interest to the reader, and so on So, it is essential to select the important information and neglect the other contents. Content selection is simply a classification task that selects important information and neglects other informations.
- Information sorting: If a summarization system selects the essential information contents, it must be sorted accordingly. The summary should not deviate from the actual meaning of the existing content. Systematic ordering might be possible for sole or unidocument text summarization but it is not applicable for multidocument text summarization [9].
- Sentence generation: Generating a sentence from the sorted information is an important process that makes the summarized information to be readable for the user. The result must be fluent and relatively express the actual information.
- Complexity: In NLP, process of text summarization is key issue since the extraction of meaningful information from a text is a complex process. Various semantics, interpretations, and dimensionalities [10] are needed to be understood which increases the system complexity.
- Word uncertainty: If the text has abbreviations, it will create uncertainty if the abbreviation has diverse meanings. It is a must for a summarization model to match the acronym relevant to the subject.
- Evaluation of summary: Developing a standard system to validate the summary is one difficulty that exists in text summarization modules. The generated results

must be compared with the standard system but it is difficult to categorize what is the correct summary.
- Summarization quality: User preference might vary from one over the other since some users might be interested in the body of the text while others might be interested in other portions. This results in quality variation in the summary.
- Reduction rate: High reduction rate must be achieved. However, it is possible for a single document text summarization and complex for multiple document summarization process. Without high background knowledge, it is complex to achieve a high reduction rate [11].

Research Contribution

This research work has considered the following contributions to get better performance for the abstractive text summarization process.
- To improve the summarization of document and multisentence representation for better extraction of important features from the given document. A novel encoding algorithm is used to improve representations that obtain a better summary.
- Sentence clustering is an important process for abstractive text summarization. To improve the sentence clustering process, adaptive Gaussian mixer model is introduced. It helps to produce better clustering results to improve the summarization process. Clustering is an effective process in the grouping of similar sub-sets based on scores of similarity measures.
- Enhanced sentence selection is preferred to improve the feature selection process effectively and removing redundant phrases. A deep belief neural network is used to select the sentences and attains better summarization results.
- Finally, improved text representation is achieved to enhance the abstractive summarization process by improving the text representation process.

Related Works

Various text summarization works are addressed in this section to obtain the research motivation. The research works are analyzed based on the techniques along with merits and demerits.

The bi-stage method for variable-length abstractive summarization reported by Ming-Hsiang et al. [12] presented a transformer-based summarization model for abstractive summarization. Using pretrained bidirectional encoder representation from transformer the text segmentation is performed and bidirectional LSTM is used in the summarization module to split the input into segments. Finally, a bidirectional encoder representation from a transformer-based model is developed to gather the essential informative sentences from every segment.

Multifeature and fuzzy-based text summarization are reported by Darshna Patel et al. [13] and Yan Du and Hua Huo [14] for keyword-based text summarization. The important features are extracted initially on the basis of predefined scores. If the score is equal to or above the predefined scores, then those important words are considered first. Then using a feature is mixed in the form of linear fashion, for identification of the vital sentences and weighted using a genetic algorithm. This improves the sentence feature selection process. Finally, fuzzy logic is employed to obtain the final score for automatic text summarization.

The double attention pointer network-based encoder and decoder model is proposed by Zhixin et al. [15] for text summarization. The essential information from the encoder is collected and more coherence contents are generated using the pointer network. Research work efficiently handles the repetition problem. Further summarization model is optimized using scheduled sampling and reinforcement learning. Reduced repetition occurrence and enhanced summarization performance are the major advantages observed in the research model.

The extended S2S neural text summarization model is presented by Jingwei et al. [16]. The extended model utilizes rule-based augmentation encoder and thematic line aware decoder to improve the summarization performance. Synthetic units are considered for the summarization process through hierarchical attention mechanism and decoder performance is improved using dual memory cell LSTM. The enhanced model achieved high performance in text summarization by generating high-quality sentences with high reliability and less redundant.

Recurrent neural network (RNN)-based S2S attentional model is reported in [17, 18] research models to prove the effective performance of abstractive text summarization. A double encoding model is preferred in the research work. Coarse encoding is attained through the primary encoder and fine encoding is performed in the secondary encoder to generate important words from the raw input text. Two encoder outputs are combined and processed in the conversion unit to obtain a assorted summary that has less replication of features.

Multistage clustering for unstructured text documents summarization is reported by Muhammad Yahya Saeed et al. [19]. It combines metadata with extracted text to improve the performance of the text summarization model. Conventional data processing, developing of variety of corpuses, initial level clustering, creating candidate subset of corpuses, making association between subset of corpuses, creating and ranking of keyword dictionary for every subset of corpus, final level clustering, creation of edge association among clusters for subset of corpuses, and processing of texts by both forward and backward propagation to generate results are the ten steps followed in the summarization model. It outperforms in all aspects compared to the conventional methods.

Graph-based automatic text summarization model is reported by Abeer Alzuhair and Mohammed Al-Dhelaan [20] which reduces the limitations present in ranking and weighting methods. Several types of graph-based approaches are collectively used to enhance the performance of summarization which makes the application suitable for natural language processing. The weighting schemes and ranking methods are combined as a single approach considering the arithmetic progression average and the harmonic progression average to attain better summarization performance.

Along with the above recent research works, some of the important research models are analyzed and observations are presented in Table 2.

Motivation and Objective

The selected features are considered as optimal parameters and top sentences are selected using the topological sort algorithm. Feedforward artificial RNN with a autonomy-based attention mechanism is defined to encode the keywords for top sentences and it is decoded to formulate the abstract text summary. The feature characteristics of RNN eliminates the computational limitations of natural language processing summarization techniques and provides better results which are similar to summary of a human. Moreover, RNN can able to handle large data and produce better accuracy compared to other models. For this benefit the proposed model is designed using the RNN model. The process flow of the proposed abstractive text summarization model is depicted in Figure 2.

To extract the top sentences from the article, all the articles are preprocessed. In the preprocessing, paragraphs are converted into sentences, special characters and stop words are removed sentences with unique meaning are identified by the process of tokenization. Tokenization is basically concerned with fragmenting the given text into multiple smaller parts consisting of words essentially named as tokens. Lemmatization is also a critical stage of the preprocessing stage to deal with inflected words. Lemmatization and stemming go together. Sentence weight frequency is obtained based on the word repetition and replace the words with weight values. Finally, it is progressed for the sentence clustering process.

Sentence clustering is introduced in the proposed approach to consolidate similar information as a single source to reduce the complexities in abstractive text summarization. Processing the same information will increase the computation time. Conventional Gaussian model makes only assumption that restricts the number of distribution to represent a cluster. Moreover, GMM tends to group data points that belong to a single distribution together. In the proposed work an adaptive Gaussian mixture model is utilized. In the adaptive approach, the model learns the system and updates the parameters for each arrival of index terms. In the clustering approach, the learning ability helps to cluster the documents efficiently. Figure 3 depicts the sentence clustering process model using the adaptive Gaussian mixture model.

Table 2: Observations from existing techniques in text summarization.

Author	Technique	Overview	Merits/demerits
Ming-Hsiang et al. [12]	Bidirectional encoder representation from transformer-based summarization model	Pretrained BEFT for text segmentation and BiLSTM are used to split the input into segments	Improved accuracy of 70% in human subject evaluation
Yan Du and Hua Huo [14]	Multifeature and fuzzy-based text summarization	The combination of features is used to extract the vital sentences and weighted using genetic algorithm, feature selection using fuzzy logic	Better performance than MSword, System 19, System 21, System 31, SDS-NNGA, GCD, SOM models
Zhixin Li et al. [15]	Double attention pointer network-based encoder and decoder model	Essential information from the encoder is collected and more coherence contents are generated using the pointer network. Scheduled sampling and reinforcement learning for optimization performance	Reduced repetition occurrence and enhanced summarization performance
Jingwei et al. [16]	Extended S2S neural text summarization model	Rule-based augmentation encoder and thematic line aware decoder to improve the summarization performance. Decoder performance is improved using dual memory cell LSTM	High reliability and less redundant
Kaichun et al. [17]	Recurrent neural network-based S2S attentional model	Coarse encoding is attained through primary encoder and fine encoding is performed in secondary encoder to generate important words from the raw input text.	Less repetition of features
Muhammad Yahya Saeed et al. [19]	Multistage clustering for unstructured text documents summarization	Clustering approaches followed by forward propagation model in conjunction with backward propoagation model are utilized for summary generation.	Better summarization performance

AbeerAlzuhair and Mohammed Al-Dhelaan [20]	Graph-based automatic text summarization model	Four weighting schemes and two ranking methods are combined as a single approach	Reduces the limitations present in ranking and weighting methods
Qian Guo et al. [21]	MS-Pointer Network that based on the multihead self-attention mechanism	Multihead self-attention to combine the semantics	Out of vocabulary issue is reduced
Deepak Sahoo et al. [22]	Abstract text summarization	Markov clustering principle is used to obtain sentence level relationship and classification technique is used to generate the abstract summary	Improved summarization performance among cluster models
Nabil Alami et al. [23]	Automatic text summarization	Word to vectors and unlabeled classification learning methods are combined for information merging from different sources	Reduce high dimensional and sparse data issues
Hadrien et al. [24]	Query-oriented text summarization	Fuzzy hypergraph approach to define the semantic connection among sentences	Eliminates associated optimization problem
Jiyuan Zheng et al. [25]	Abstractive text summarization	Organized sets of adaptive sequential encoder–decoder network learning to interpret the actual meaning among representation of conversations	Improved summarization performance
Mahmood Yousefi-Azar and Len Hamey [26]	Query-oriented single-document summarization	Encoders and decoders to compute a feature space from the term-frequency input	Improved performance to grab informative sentences selection
Razieh et al. [27]	Automatic text summarization	The granular-based automata is defined to interpet sentence extraction, an artificial bee colony optimizes the fitness function, and text feature weights are adjusted using PSO and GA	Classifies more important and less important features effectively

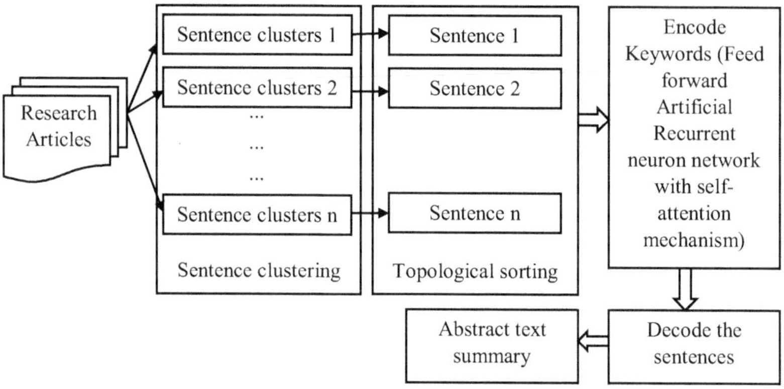

Figure 2: Process flow of proposed abstractive text summarization model.

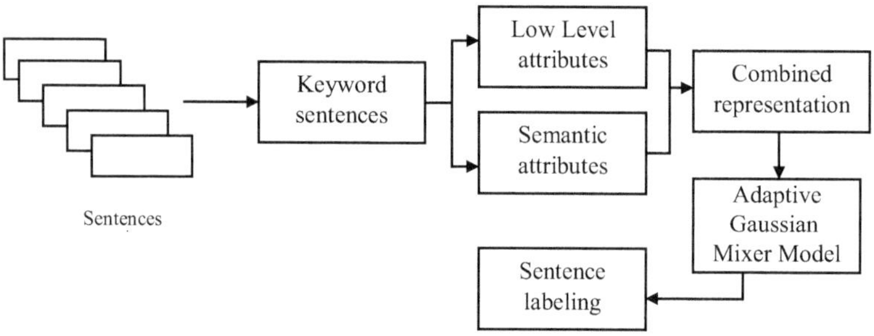

Figure 3: Sentence clustering process model using adaptive Gaussian mixture model.

Consider at time t, a new index term is identified, the numerical representation of the new index is extracted, and its parameters are updated using the Gaussian mixture model. The maximum number of possibilities for Gaussian is assumed as k. Each Gaussian represents one of the possible index distributions. This defines the sentence class as labels and it is grouped as clusters. The time for clustering the sentences depends on the learning rate. If the learning rate is low, processing speed is reduced but at the cost of increasing time complexity. On the other hand, in the case of high learning rates, the changes are updated more frequently. For better extraction, a low learning rate is preferred in the research model. The labeling is performed by calculation of affinity-based distance similarity among the sentences. To define the similarity distance, Gaussian equation is derived as follows.

$$\Delta\left(\vec{i}_t, i\right) = \sqrt{\left(\vec{i}_t - \vec{\mu}_{t,i}\right)^T \cdot \sum_{i,t}^{-1} \cdot \left(\vec{i}_t - \vec{\mu}_{t,i}\right)} \tag{1}$$

where $\vec{i}_t$ is the sample vector at time t. The mean function is represented as $\vec{\mu}_{t,i}$ and covariance is represented as $\sum_{i,t}^{-1}$. The distance threshold is defined to add the sentences to the clusters and the deviation from the average is defined as a standard deviation which is a square root of the average of all the squared deviations. One standard deviation is termed as one sigma which includes 68% of all the data points. But generally, 2.5σ is used as the threshold value which included 99.7% of Gaussian distribution data points. Depends on the feature length the threshold range can be varied. The weight update for all the Gaussian is derived as

$$W_{i,t} = (1 - \alpha) W_{i,t-1} + \alpha(m_{i,t}) \tag{2}$$

where the value of $m_{i,t}$ is 1 if the sentences are matched; otherwise, it is considered as zero.

Once the sentences are clustered, top sentences are identified and sorted using the topological sorting approach. Topological sorting is one of the standard algorithms for ordering the sentences as directed graphs. Sorting results presents the ordered vertices for every directed edge. For every vertex, the sentences are arbitrarily moved into a new position. Initially, each node in the graph is looped using a depth search-based algorithm and analyzes the sentences outgoing edges. The ordering between the sentences is identified and represented as top sentences by defining the distance between the directed graph and directed edges. Figure 4 gives an illustration for topological sorting used in the proposed text summarization process.

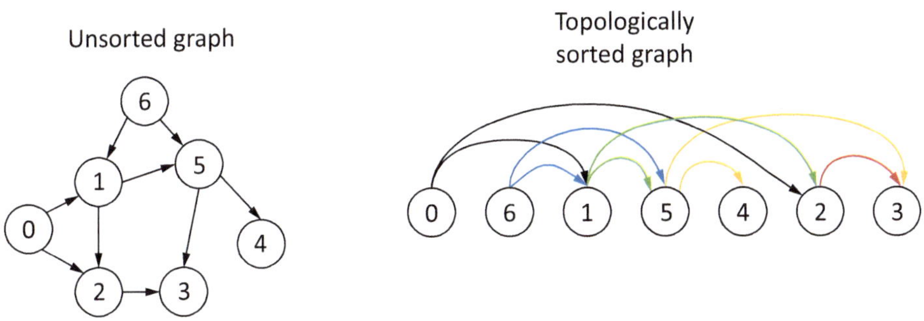

Figure 4: Topological clustering.

After sentence clustering and topological sorting, the keywords are encoded to extract useful information from the large volume of texts. Artificial RNN with self-attention mechanism is proposed in this research work for efficient keyword encoding. The mechanism introduced here involves essential features selection such as keywords and ignore the other elements in the sentence. In the encoding phase, the entire input sequence is processed to obtain a context vector which is obtained in the last stage of the RNN model. This final stage encoder output is provided as input to the decoder while ignoring other encoder outputs. The decoder RNN unit produces the words in a sentence sequentially. A simple scheme of the RNN-based encoding–decoding process is illustrated in Figure 5.

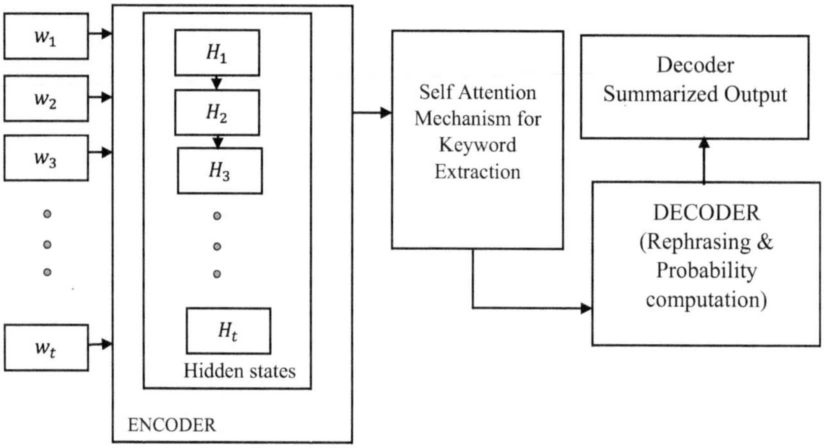

Figure 5: Scheme of proposed RNN-based abstractive text summarization.

Encoder acquires the input sequence and produces vector sequence $v = \{v_1, v_2, \ldots, v_{t_v}\}$, which is then converted into a single vector v^2. The essential condition to utilize RNN for the encoding process is expressed as

$$H_t = f(v_t,\ H_{t-1}) \tag{3}$$

$$v_g = k([H_1, H_2, \ldots, H_{v_t}]) \tag{4}$$

where the hidden states are represented as H_t at time t. The vector generated from the hidden states sequence is given as v_g. The terms f and k represent the nonlinear functions. The encoder functions based on a feedforward neural network with directionality from input toward the output side. As mentioned in previous stages, the context vector is generated from a list of keyterms through the own autonomy-based attention mechanism. The autonomy-based attention mechanism is an analogous concept wherein the reader is caught attention or interest toward certain terms while reading a sentence or a paragraph. These terms or keywords might increase the impact of model's attention towards specific topical words that defines best event described in those texts. The context vector which is a collection of these keywords generated from the self-attention system is concatenated. The context vector is given to the decoder. Beforehand, the decoder is initialized with the last state of the encoding model. It is to be noted that local attention mechanisms are used for the formulation of context vectors where only a selected set of hidden state information is used to generate the keywords. This accounts for the observed reduction in vector dimension consequently causing a reduced time complexity.

The decoder RNN stage dealt with the word predictions from the contextual vectors using all the predefined set of values. START and END tokens initialize and end the learning process in the decoder module. The decoder is the module where rephrasing of the sentence in an abstractive form from the set of self-attention-based keywords is done. The pseudocode for the encoding and decoding phase could be illustrated as follows. Initially, in the encoding phase the preprocessing steps like tokenization, lemmatization, and stemming are performed and terms are labeled. Upon updating the weight function the sentence matching process is initiated and assigns values as 1 for matched sentence and 0 for other sentences. Finally, the vector sequence and hidden states are used to generate the vectors for every sentence in the document. The Decoding phase deals with regeneration of sentences based on the vector length specifications and context the sentences are rephrased and weight functions are generated for the encoded inputs.

Pseudocode for the encoding phase

```
start
initalize preprocessing tokenization, lemmatization, stemming
initialize labelling using eqn. (1)
update weight W_{i,t}
{
                begin sentence matching
{
            Assign (m_{i,t} = = 1) for matched sentence
        else assign (m_{i,t} = = 0)
        }
{
call * / vector sequence /*
Intialize hidden states H_t
generate vectors v_g for each sentences
repeat
}
end
```

Pseudocode for the encoding phase

```
start
initalize decode_initial < - encloder_last
initialize context vector length, context vector c_v (i)
while token (i(0)) = = START
begin
{
call rephrase * / context vector /*
check P(w_max) for each c_v (i)
generate W_new for every P(w_ max = = true)
else
c_v(i + 1)
}
repeat until i(w_t) = = END
```

The probability over the translation is achieved by sequentially decomposing the joint probability and it is given as

$$P(w) = \prod_{t=1}^{T} P\left(w_t \middle| [w_1, w_2, \ldots, w_{t-1}], v_g\right) \tag{5}$$

where $w = [w_1, w_2, \ldots, w_{T_w}]$ and the conditional probability for RNN is derived as

$$P\left(w_t \middle| [w_1, w_2, \ldots, w_{t-1}], v_g\right) = l\left(w_{t-1},\ H_s, v_g\right) \tag{6}$$

where l is a nonlinear function which is the multilayered probability function of w_t, and the hidden state of RNN is represented as H_s. The decoded text provides the summarized features from the large documents. The limitations of the conventional RNN model are reduced due to the initial sentence clustering and sorting process. The selection of top sentences has reduced the long sentence processing issue in the traditional RNN model. The novelty of the research work is present in its hybrid approach which combines multiple methodologies such as adaptive Gaussian mixture and topological sorting along with the FFRNN. In the classical methods, text summarization is achieved through classification approaches which do not concentrate much on preprocessing steps. This makes the conventional systems lag in performance while handling large phrases and this issue is overcome by the proposed approach.

Results and Discussion

The abstractive text summarization proposed model is validated by analyzing the experimental results. Three datasets that contain abstracts of research articles are considered to evaluate the system performance. NSF dataset contains NSF research abstracts, NIPS dataset contains NIPS papers, and AAN dataset has abstracts of AAN. In the NSF dataset, abstracts from diverse scientific research articles from 1990 to 2003 are considered. In which 10 years of data from 1990 is considered for the training process and articles from 2001 to 2003 are considered for testing and the documents of the year 2000 are considered for validation. In the NIPS dataset, 10 years of data from 2005 to 2015 are considered in which from 2005 to 2013 the abstracts are considered for the training process, and the documents of the year 2015 are considered for testing and 2014 is considered for the validation process. In the ANN database, 10 years of data from 2003 to 2013 is considered in which from 2003 to 2010 the abstracts are considered for the training process, and the documents of the year 2011 are considered for testing and 2012 to 2013 is considered for the validation process. The maximum length of the paraphrase sentence from NSF and ANN datasets is 30 sentences and 25 sentences for the NIPS dataset.

Dataset is preprocessed as per the steps discussed in the previous section and the performance is evaluated on the basis of matching score, accuracy of sentence, Rouge-1 which defines distance between predicted order and correct order, Rouge-s which defines sum of number of pairs in a document, and Rouge-L defines the longest common sequence. Attention-based network (ABN) and BERT-based network (BEN) are considered along with conventional RNN (CRNN) to validate the performance of the proposed FFRNN. Essential formulations utilized to obtain the evaluation parameters are given as follows:

$$\text{Matching score} = \frac{1}{N}\sum_{i=1}^{N}\left(\hat{o}^i = o*^i\right) \tag{7}$$

$$\text{Sentence accuracy} = \frac{1}{N}\sum_{i=1}^{N}\frac{1}{v_i}\sum_{j=1}^{v_i}\left(\hat{o}^i = o*^i\right) \tag{8}$$

$$\text{Rouge} - 1 = 1 - \frac{2I}{v_i^2} \tag{9}$$

$$\text{Rouge} - s = \frac{1}{v_i^2}(\hat{o} \cap o*) \tag{10}$$

$$\text{Rouge} - L = \frac{(1+\beta^2)(\hat{o},\ o*)}{\beta^2 v_i^2} \tag{11}$$

where β controls the relative accuracy. The proposed feedforward network is trained with initial learning rate of 4×10^{-4} for encoding the sentence and paragraph, and the learning rate of 4×10^{-3} is used for decoding the sentence and paragraph. For the loss function, the margin hyperparameter value γ is set into 1. The comparison of obtained matching scores for all the four models is depicted in Figure 6. The percentage of samples obtained for the entire document is observed as matching scores. It is observed from the figure that proposed hybrid approach obtains a maximum matching score which reflects the better performance of the proposed approach. The performance of conventional models reduced due to encoding and decoding the entire sentence. But in the case of the proposed approach, keywords are encoded and decoded which greatly reduces the complexity and this increases the matching score.

In Figure 7, the accuracy of the proposed approach is compared with the conventional approaches. Accuracy defines the percentage of the sentence correctly summarized by the system. It is observed that the proposed approach attains maximum gain for all three datasets by extracting the important sentences. The limitations in the conventional RNN model are overcome by the proposed approach by utilizing adaptive sentence clustering and sorting. This additional process improves the accuracy of the proposed abstractive text summarization approach.

Figure 8 depicts the values obtained between extracted order and correct order-based. Distance is calculated and presented as Rouge 1 in the figure. Observation

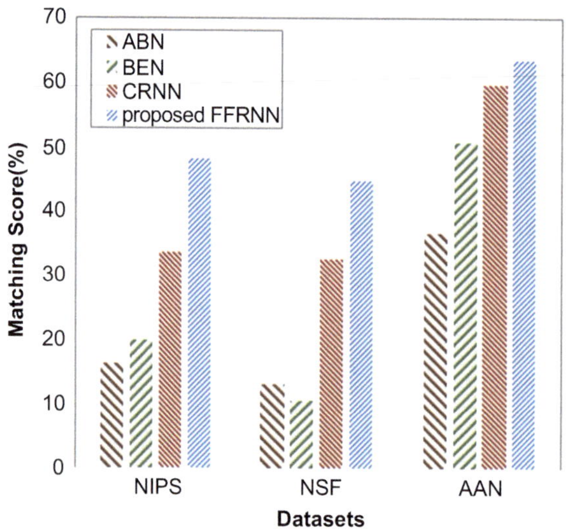

Figure 6: Matching score comparison.

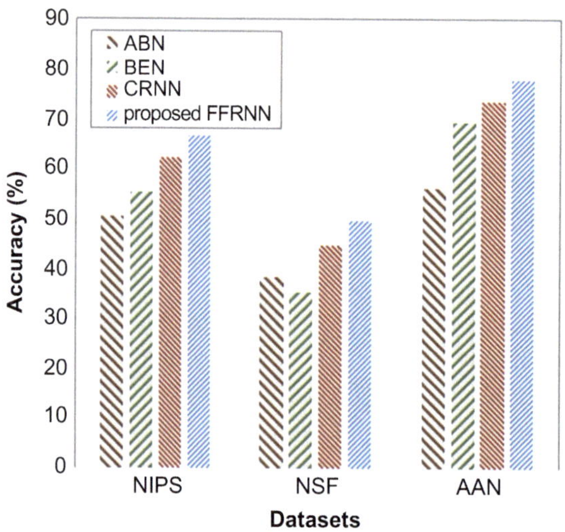

Figure 7: Sentence accuracy comparison.

from the proposed approach has the maximum distance between extracted and correct order which reflects the unique characteristics of summarization. If the distance is less the extracted content is nearly similar to actual content due to this reason the ABN performance gets reduced.

The total number of pairs in a sentence is represented as Rouge-S and depicted in Figure 9. Similar to Rouge-1, the performance of the proposed FFRNN model

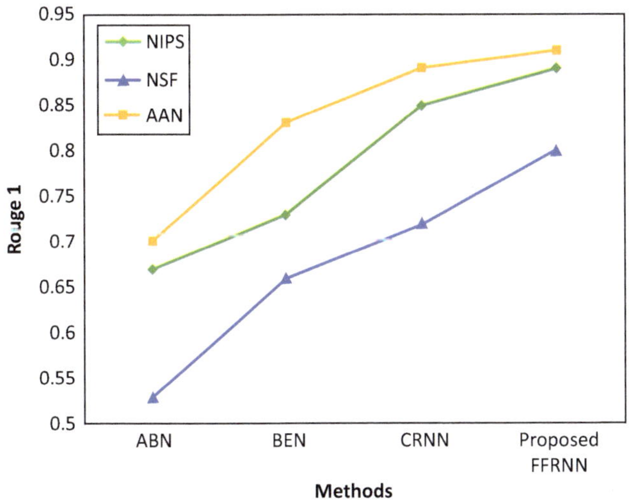

Figure 8: Rouge-1 comparison.

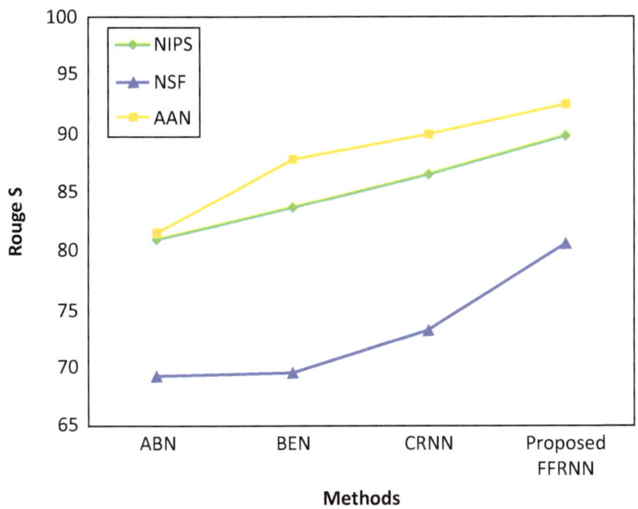

Figure 9: Rouge-S comparison.

attains a better score due to the improved clustering and sorting process performed before fed into the RNN model. The performance of ABN is almost the same for AAN and NSF datasets. However, in the last, the performance of the proposed approach reaches maximum which is 15% greater than the ABN model.

The percentage of the longest common sentence occurrence while performing abstractive text summarization is observed and plotted in Figure 10. All four models have more number of longest common sentences. However, the performance of the proposed FFRNN model has maximum common sentences which reflects the efficient text processing ability. Rather than selecting few sentences, considering all the important feature-based sentence extraction results in a large number of longest common sentences.

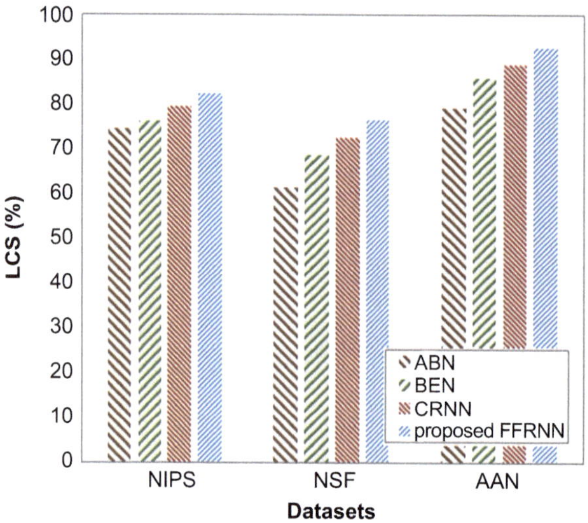

Figure 10: Longest common sentence identification.

Figure 11 depicts the comparative analysis of rouge scores for the three datasets. It is observed that the score varies for each data in different aspects. Among the three datasets, ANN has the maximum rouge scores due to simple abstract summaries.

The proposed model and conventional models performance metrics are listed in Table 3 for matching score and efficiency. From the results, it is observed that the proposed FFRNN model attains better performance compared to conventional methods.

The efficiency of the proposed abstractive summarization model is calculated based on the matching score and it is depicted in Figure 12. It is observed that the proposed FFRNN model attains maximum efficiency compared to other models. This improved performance makes the system suitable for various application domain which requires abstractive text summarization.

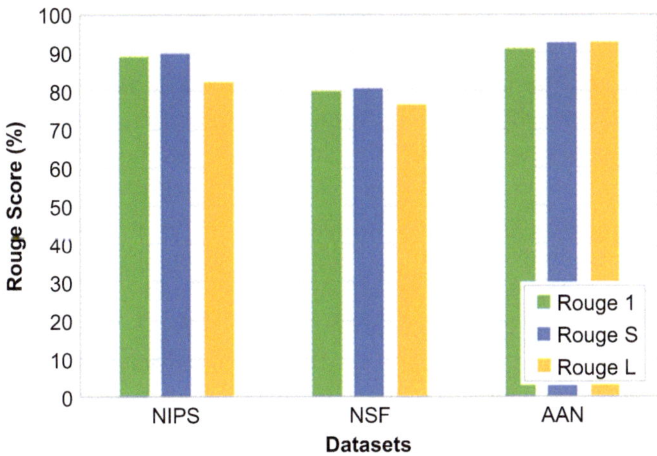

Figure 11: Comparison of rouge scores.

Table 3: Performance metric comparison.

Methods	Matching score (%)			Accuracy (%)		
	NIPS	**NSF**	**AAN**	**NIPS**	**NSF**	**AAN**
ABN	16.25	13.18	36.62	50.5	38.28	56.22
BEN	19.9	10.44	50.76	55.23	35.21	69.22
CRNN	33.54	32.65	59.64	62.24	44.68	73.54
Proposed FFRNN	48.21	44.82	63.55	66.48	49.58	77.82

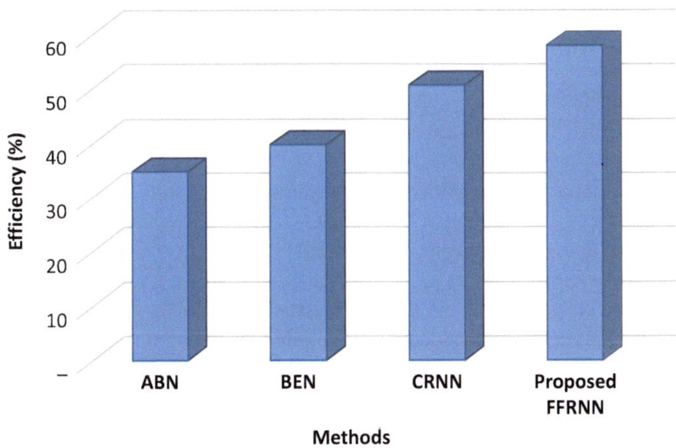

Figure 12: Efficiency comparison.

Conclusion

A novel deep belief neural network for abstractive text summarization is presented in this research work. Conventional abstractive text summarization models lag in performance while handling large sentences. To overcome this, feedforward artificial RNN with self-attention model is proposed in the research work which attains better performance than the conventional model. This hybrid approach is obtained by including adaptive Gaussian mixture model for sentence clustering which categorizes similar sentences, and topological sorting is used to compute the top sentences. Encoding and decoding stages are processed through FFRNN. Experimental result demonstrates the improved performance of proposed abstractive text summarization model than conventional models. Though the proposed approach attains better accuracy than existing techniques it needs to be improved which is considered as the limitation of the proposed work. Further, this research work can be improved by introducing a hybrid deep learning model to improve the matching performance and accuracy.

References

[1] D. S. Moirangthem and M. Lee, "Abstractive summarization of long texts by representing multiple compositionalities with temporal hierarchical pointer generator network," Neural Networks, vol. 124, pp. 1–11, 2019.

[2] R. Mishra and T. Gayen, "Automatic lossless-summarization of news articles with abstract meaning representation," Procedia Computer Science, vol. 135, pp. 178–185, 2018.

[3] F. Kiyoumarsi, "Evaluation of automatic text summarizations based on human summaries," Procedia – Social and Behavioral Sciences, vol. 192, pp. 83–91, 2015.

[4] L. Ermakova, J. V. Cossu, and J. Mothe, "A survey on evaluation of summarization methods," Information Processing & Management, vol. 56, no. 5, pp. 1794–1814, 2019.

[5] J. V. Tohalino and D. R. Amancio, "Extractive multi-document summarization using multilayer networks," Physica A: Statistical Mechanics and Its Applications, vol. 503, pp. 526–539, 2018.

[6] S. Gupta and S. K. Gupta, "Abstractive summarization: An overview of the state of the art," Expert Systems with Applications, vol. 121, pp. 49–65, 2019.

[7] M. J. Mohan, C. Sunitha, and A. Jaya, "A study on ontology based abstractive summarization," Procedia Computer Science, vol. 87, pp. 32–37, 2016.

[8] H. Oufaida, O. Nouali, and P. Blache, "Minimum redundancy and maximum relevance for single and multi-document Arabic text summarization," Journal of King Saud University – Computer and Information Sciences, vol. 26, no. 4, pp. 450–461, 2014.

[9] L. Marujo, W. Ling, and J. P. Neto, "Exploring events and distributed representations of text in multi-document summarization," Knowledge-Based Systems, vol. 94, pp. 33–42, 2016.

[10] P. Janaszkiewicz and P. Różewski, "The method of multidimensional approach to text summarization," Procedia Computer Science, vol. 159, pp. 2189–2196, 2019.

[11] J. Zhang, K. Li, and C. Yao, "Event-based summarization for scientific literature in Chinese," Procedia Computer Science, vol. 129, pp. 88–92, 2018.

[12] S. Ming-Hsiang, W. Chung-Hsien, and H.-T. Cheng, "A two-stage transformer-based approach for variable-length abstractive summarization," IEEE/ACM Transactions on Audio, Speech, and Language Processing, vol. 28, pp. 2061–2072, 2020.
[13] D. Patel, S. Shah, and H. Chhinkaniwala, "Fuzzy logic based multi document summarization with improved sentence scoring and redundancy removal technique," Expert Systems with Applications, vol. 134, pp. 167–177, 2019.
[14] Y. Du and H. Huo, "News text summarization based on multi-feature and fuzzy logic," IEEE Access, vol. 8, pp. 140261–140272, 2020.
[15] Z. Li, Z. Peng, S. Tang, C. Zhang, and H. Ma, "Text summarization method based on double attention pointer network," IEEE Access, vol. 8, pp. 11279–11288, 2020.
[16] J. Cheng, F. Zhang, and X. Guo, "A syntax-augmented and headline-aware neural text summarization method," IEEE Access, vol. 8, pp. 218360–218371, 2020.
[17] K. Yao, L. Zhang, D. Dawei, T. Luo, L. Tao, and Y. Wu, "Dual encoding for abstractive text summarization," IEEE Transactions on Cybernetics, vol. 50, no. 3, pp. 985–996, 2020.
[18] R. Adelia, S. Suyanto, and U. N. Wisesty, "Indonesian abstractive text summarization using bidirectional gated recurrent unit," Procedia Computer Science, vol. 157, pp. 581–588, 2019.
[19] M. Y. Saeed, M. Awais, R. Talib, and M. Younas, "Unstructured text documents summarization with multi-stage clustering," IEEE Access, vol. 8, pp. 212838–212854, 2020.
[20] A. Alzuhair and M. Al-Dhelaan., "An approach for combining multiple weighting schemes and ranking methods in graph-based multi-document summarization," IEEE Access, vol. 7, pp. 120375–120386, 2019.
[21] Q. Guo, J. Huang, N. Xiong, and P. Wang, "MS-Pointer network: Abstractive text summary based on multi-head self-attention," IEEE Access, vol. 7, pp. 138603–138613, 2019.
[22] D. Sahoo, A. Bhoi, and R. C. Balabantaray, "Hybrid approach to abstractive summarization," Procedia Computer Science, vol. 132, pp. 1228–1237, 2018.
[23] N. Alami, M. Meknassi, and N. En-nahnahi, "Enhancing unsupervised neural networks based text summarization with word embedding and ensemble learning," Expert Systems with Applications, vol. 123, pp. 195–211, 2019.
[24] H. Van Lierde and T. W. S. Chow, "Learning with fuzzy hypergraphs: A topical approach to query-oriented text summarization," Information Sciences, vol. 496, pp. 212–224, 2019.
[25] J. Zheng, Z. Zhao, and X. Yan, "Abstractive meeting summarization by hierarchical adaptive segmental network learning with multiple revising steps," Neurocomputing, vol. 378, pp. 179–188, 2019.
[26] M. Yousefi-Azar and L. Hamey, "Text summarization using unsupervised deep learning," Expert Systems with Applications, vol. 68, pp. 93–105, 2016.
[27] R. Abbasi-ghalehtaki, H. Khotanlou, and M. Esmaeilpour, "Fuzzy evolutionary cellular learning automata model for text summarization," Swarm and Evolutionary Computation, vol. 30, pp. 11–26, 2016.

Index

https://doi.org/10.1515/9783110750584-012

De Gruyter Series
Smart Computing Applications

ISSN 2700-6239
e-ISSN 2700-6247

Cloud Analytics for Industry 4.0
Edited by Prasenjit Chatterjee, Dilbagh Panchal, Dragan Pamucar and Sharfaraz Ashemkhani Zolfani, 2022
ISBN 978-3-11-077149-7, e-ISBN (PDF) 978-3-11-077157-2,
e-ISBN (EPUB) 978-3-11-077166-4

Advances in Industry 4.0
Concepts and Applications
Edited by M. Niranjanamurthy, Sheng-Lung Peng, E. Naresh, S. R. Jayasimha, Valentina Emilia Balas, 2022
ISBN 978-3-11-072536-0, e-ISBN (PDF) 978-3-11-072549-0,
e-ISBN (EPUB) 978-3-11-072553-7

Soft Computing and Optimization Techniques for Sustainable Agriculture
Debesh Mishra, Suchismita Satapathy, Prasenjit Chatterjee, 2022
ISBN 978-3-11-074495-8, e-ISBN (PDF) 978-3-11-074536-8,
e-ISBN (EPUB) 978-3-11-074537-5

Knowledge Engineering for Modern Information Systems
Methods, Models and Tools
Edited by Anand Sharma, Sandeep Kautish, Prateek Agrawal,
Vishu Madaan,
Charu Gupta, Saurav Nanda, 2021
ISBN 978-3-11-071316-9, e-ISBN (PDF) 978-3-11-071363-3,
e-ISBN (EPUB) 978-3-11-071369-5

Knowledge Management and Web 3.0
Next Generation Business Models
Edited by Sandeep Kautish, Deepmala Singh, Zdzlslaw Polkowski,
Alka Mayura,
Mary Jeyanthi, 2021
ISBN 978-3-11-072264-2, e-ISBN (PDF) 978-3-11-072278-9,
e-ISBN (EPUB) 978-3-11-072293-2